EARTH SCIENCE SUCCESS

55

TABLET-READY, NOTEBOOK-BASED LESSONS

EARTH SCIENCE SUCCESS

TABLET-READY, NOTEBOOK-BASED LESSONS

CATHERINE OATES-BOCKENSTEDT
MICHAEL D. OATES

Arlington, Virginia

Claire Reinburg, Director
Wendy Rubin, Managing Editor
Andrew Cooke, Senior Editor
Amanda O'Brien, Associate Editor
Donna Yudkin, Book Acquisitions Coordinator

ART AND DESIGN
Will Thomas Jr., Director
Joe Butera, Senior Graphic Designer, cover and interior design

PRINTING AND PRODUCTION
Catherine Lorrain, Director

NATIONAL SCIENCE TEACHERS ASSOCIATION
David L. Evans, Executive Director
David Beacom, Publisher

1840 Wilson Blvd., Arlington, VA 22201
www.nsta.org/store
For customer service inquiries, please call 800-277-5300.

18 17 16 15 4 3 2 1

NSTA is committed to publishing material that promotes the best in inquiry-based science education. However, conditions of actual use may vary, and the safety procedures and practices described in this book are intended to serve only as a guide. Additional precautionary measures may be required. NSTA and the authors do not warrant or represent that the procedures and practices in this book meet any safety code or standard of federal, state, or local regulations. NSTA and the authors disclaim any liability for personal injury or damage to property arising out of or relating to the use of this book, including any of the recommendations, instructions, or materials contained therein.

Library of Congress Cataloging-in-Publication Data

Oates-Bockenstedt, Catherine.
Earth science success : 55 tablet-ready, notebook-based lessons / Catherine Oates-Bockenstedt and Michael Oates. —2nd edition.
pages cm
Includes bibliographical references.
ISBN 978-1-941316-16-0
1. Earth sciences--Study and teaching (Middle school)—United States. 2. Lesson planning—United States. I. Oates, Michael. II. Title.
QE47.A1O28 2015
550.71'273—dc23
2015001204

Cataloging-in-Publication Data for the e-book are also available from the Library of Congress.
e-LCCN: 2015001714

CONTENTS

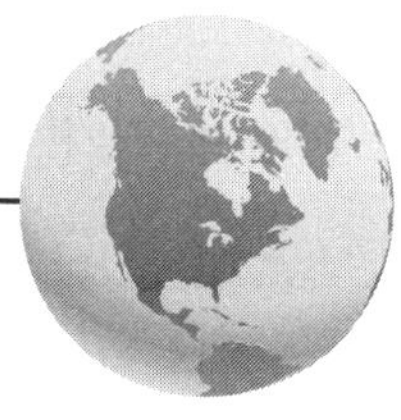

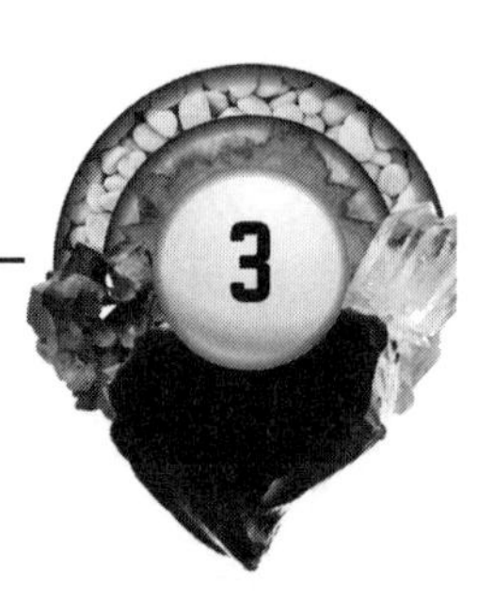

CHAPTER 3

EARTH'S SURFACE PROCESSES 109

CHAPTER 4

HISTORY OF PLANET EARTH 149

CHAPTER 5

EARTH'S INTERIOR SYSTEMS 187

CHAPTER 6

EARTH'S WEATHER 219

6

CHAPTER 7

HUMAN IMPACTS ON EARTH SYSTEMS 277

7

APPENDIXES

INDEX 319

**"THE ART OF TEACHING
IS THE ART OF ASSISTING DISCOVERY."**

MARK VAN DOREN (1894-1972)

This book is dedicated to the power that
collaboration has among classroom teachers.
Special mention goes to my friend, Mary Gallus,
for her enhancement of lessons and
expert collaboration.

INTRODUCTION

This second edition of *Earth Science Success: 55 Tablet-Ready, Notebook-Based Lessons* provides a one-year Earth Science curriculum with 55 classroom-proven lessons designed to follow the disciplinary core ideas for middle school Earth and space science from the *Next Generation Science Standards* (*NGSS*). Intended for teachers of grades 5–9, *Earth Science Success* emphasizes hands-on, sequential experiences through which students discover important science concepts lab by lab and develop critical-thinking skills. Whereas the first edition focused more on the rationale for implementing the curriculum and the wisdom of using composition notebooks, this second edition focuses a special lens on the lessons themselves. The 55 lesson plans enable teachers to use electronic tablets, such as iPads, with best practice, field-tested methods.

Middle school Earth science teachers' days are very busy with large classes, meetings and various duties, grading and correction, class preparation, answering communications from parents, and so on. *Earth Science Success* is the result of the authors' desire to create a notebook-based, lab-focused, ready-to-use, and now tablet-ready curriculum that has been field-tested and refined for success. The authors have organized this curriculum into a series of investigations that emphasize the active involvement of students in a discovery process. Intended primarily for classroom science teachers as a survival guide for teaching a full Earth science course, *Earth Science Success* follows a three-step pattern of active involvement in the discovery process, which includes anticipation, evidence collection, and analysis. The topics chosen and the laboratory approach employed in *Earth Science Success* reflect the core ideas involved in scientific and engineering practices, which lead to the four main categories of performance expectations from *NGSS*: Engineering Design, Earth's Place in the Universe, Earth's Systems, and Earth and Human Activity. *Earth Science Success* is also a valuable tool for training future science teachers, who will enjoy implementing and discussing the investigations featured in this book.

To the Earth Science Teacher

Like you, the author is a busy classroom science teacher. Successful strategies include those that save time and promote skillful organization. Both composition notebooks and electronic tablets offer tremendous opportunities in this regard. The lessons in *Earth Science Success* lend themselves toward either approach. Combining the two, however, is even better.

The same successful pattern is followed for each lab report, no matter what the learning target or concept. See "Expectations for Each Investigation," p. XIII, for a summary of the expectations for each component of the lab report. The point value shown in parentheses is flexible, and is based on a 30-point total for each lab report grade.

Among iPad apps, *Paperport Notes*, *Evernote*, and *Notability* all provide for fully integrated note taking. The author uses the *Notability* app with the Divider set as Science, and the Subjects set as Labs and Lessons, Reference Pages, and Glossary. She creates a PDF of each lab report template and posts it on her website (both *Google* and *Schoology*), for students to download. She encourages "auto syncing" to *Google Drive* or *Dropbox*, so if a glitch happens with the electronic lesson, the work has been backed up. Students submit their assignments electronically to *Schoology*, but *Showbie* and *eBackpack* also work well in that capacity. These Learning Management Systems allow teachers to "push" the assignments onto the students' tablets, and provide due date calendar systems, as well.

The author uses a mini-conference method for typical in-class grading. This involves collecting all of the iPads (or composition notebooks with bookmarks placed in the current lab) in the front of the classroom, on a cart. The lab reports are graded in random order, while students work on other assignments and lessons, such as the graphing or analysis portions of the following lab, at their desks. Students are called up to sit next to the teacher, to witness the grading, as individuals, in a semiprivate conference. Input from the student and feedback from the teacher become clear and lasting through the use of this method. The author has found that a class of 32 iPad lab reports can be graded using the mini-conference system in a typical 50-minute class period.

Why are notebooks, both electronic and nonelectronic, so valuable? One of the most important reasons is that students are able to organize, reflect upon, and retrieve their learning. This enables them to increase their scores assessments and achieve at higher levels. Students tend to have fewer missing assignments, and "no name" papers are a thing of the past. While tablets enable connections to internet research, word-processing capabilities, real-time data, and access to rich video vignettes to expand learning, the composition notebooks have many benefits as well. Composition notebooks are durable. The fact that no pages can be torn out enables students to refer to past results. Any important handouts and foldables can be glued or taped in, and students can incorporate labeled sketches, data tables, predictions, analysis questions, personal reflection, vocabulary, and correction of misconceptions in each lesson. The tablets and composition notebooks are great

resources to use at parent/teacher conferences. The evidence to show student learning through investigations is clear. By the time students reach middle school, using hands-on activities to teach meaning in science is critical (NAEP 2013).

Expectations for Each Investigation

Based on a 30-point total, the author uses the following system to grade lab reports.

1. **Title (1 point):** The title should include several descriptive words, not a complete sentence, dealing with the chosen topic of the experiment. It should be brief and catchy, but should also indicate the variable(s) that were tested.
2. **Problem (1 point):** This provides the anticipatory question, which lends focus to learning for the experiment. It should be a complete sentence and phrased as a question. The problem explicitly states the experimental question being investigated, providing enough detail so the audience can understand what will be done.
3. **Prediction (1 point):** The prediction (or hypothesis, depending on the particular requirements in each lab) must be a complete sentence and on-topic. It is not graded for accuracy, but it is often compared and contrasted later with the final outcome of the investigation. This is where the author often targets the correction of misconceptions through class discussion and formative assessment.
4. **Thinking About the Problem (3 points):** This section gives the student necessary background information and content descriptions related to the investigation. The expectation is for the student to develop strengths in literacy by highlighting important sentences while the teacher reads the section out loud. This process helps the student to write three main points from the background information (see Figure I.1, p. XIV for an example). The teacher should have students share several main points out loud, after writing, so that misconceptions can be anticipated and explained.
5. **Labeled Image (3 points):** The image should clearly show the labeled materials and experimental setup, so that the student can describe all procedures. On each lab report, there is a designated space where students can place their image (or draw their sketch, when using composition notebooks). See Figure I.2, p. XV, for an example.

FIGURE I.1.

STUDENT SAMPLE OF "THINKING ABOUT THE PROBLEM" SECTION IN SCIENCE NOTEBOOK

Problem: What is so special about water?
Prediction: Give a working definition of water molecule.

Water Molicules will stick to their Suroundings.

Thinking about the Problem:
What does H_2O mean? Each molecule (*molecula* 'small bit' in Latin) of water is made of two hydrogen atoms (H_2) and one oxygen atom (O). What is special about water molecules is that they tend to "stick" to each other (cohesion) and to other molecules (adhesion). They do this because water is built like a magnet, with a positive end and a negative end. This helps it bond well.

Water makes life on Earth possible. It covers almost three-fourths of the surface of our planet. Because there is so much of it, water may seem very ordinary to us, and yet it is unique when compared to all other substances. For example, water is the only substance on Earth that occurs naturally in all three states-solid, liquid, and gas. In addition, solid H_2O (ice) is less dense than its liquid form (water), so it floats. Most other solids are denser than their liquid form, so they sink! Another difference, with respect to water, is that large amounts of energy must be added to water to achieve even a relatively small change in temperature. That is why our oceans moderate the temperatures of coastal communities on Earth.

Each

Thinking about the Problem:
1. Each Molecule of water is Made of two hydrogen atoms (H^2) and one oxygen atom (O).
2. water is built like a magnet, with a positive and negative end.
3. water is the only substance on Earth that occures naturally in all three States -Solid, liquid, and gass.

Materials:
8 oz. Drinking Glass
Dish Soap
Eye Dropper
A variety of water containers (assortment of five glasses, buckets, bowls, etc.)
Many Pennies (or replacement item…control for size)
Other Coins

Procedures:
1. Predict which of your five large containers (each full to the rim with water) will be able to withstand the addition of the greatest number of pennies (or replacement item) without spilling over. Test and record your results. Take photos with your iPad while conducting this step.
2. Place a dry penny on a piece of paper towel.
3. Predict the number of drops you can pile on the penny before water runs over the edge.
4. Test and record for each particular coin. Take photos with your iPad during this step.
5. Draw/phtograph and label a sketch/image of the water on the surface of the coin just before the water spilled over.
6. Conduct the same tests with the soapy water.

Analysis:
1. Describe the shape of the water as it "sits" on a coin.

Curved on the penny like a upsidedown U

water

coin

FIGURE I.2.

STUDENT SAMPLE OF LABELED IMAGES IN SCIENCE NOTEBOOK

2. Why does water pile up on a coin, rather than spilling over the edges immediately? How is the soapy water different? (Describe the science behind your thoughts…"Thinking about the Problem" will help you here.)

The water sticks making it curve up and the Soapy water does not stick so it falls of the edge.

3. Use science concepts to suggest reasons why each of the five containers holds a different number of pennies.

I found out that a glass with a bigger rimb will hold more pennys because their is more surface area/more room for water to bend around rimb.

4. Explain "surface tension" as if you were explaining it to a second grader.

Allows objects like bugs to sit on water.

Labeled Sketches/Images:

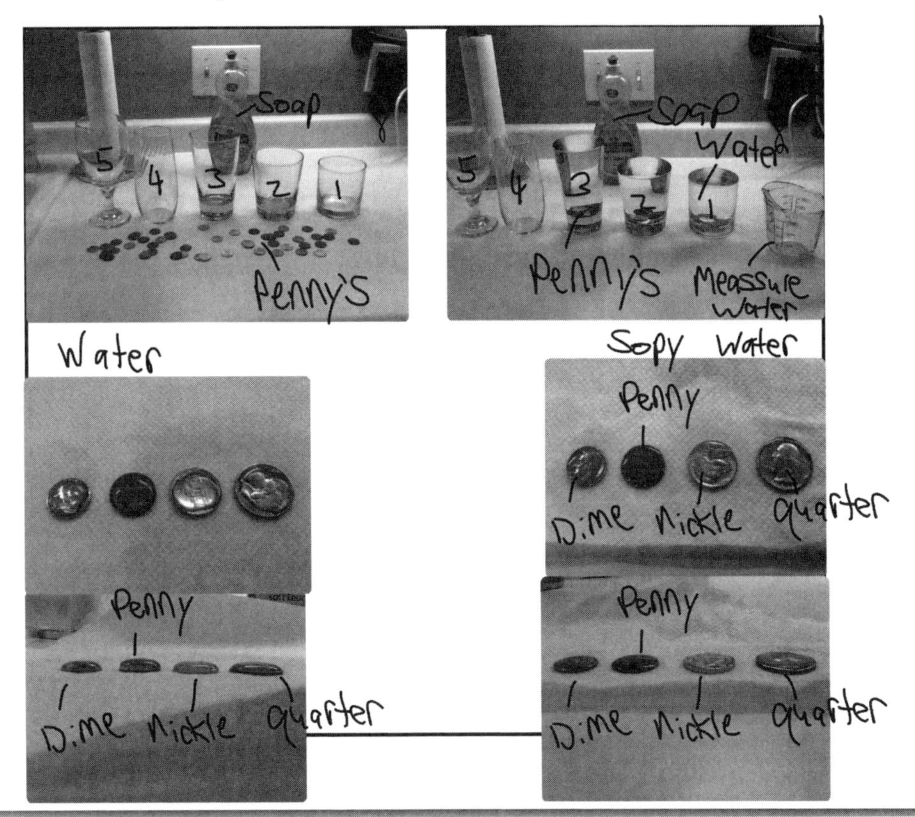

6. **Data Tables and Graphs (8 points):** All labs contain at least one data table, and many include graphs, charts, concept maps, and so on. Students are expected to correctly label graphs and tables so the audience can understand them. The evidence presented in the data tables and graphs should be complete and accurate.
7. **Analysis Questions (8 points):** These questions vary in their approaches. Some require students to describe the purpose and procedure for the experiment, while others require a discussion about how variables were controlled. Students will describe what evidence was observed and what was measured, and they will compare the original prediction with the results to help defeat misconceptions. They will also include how the results are supported by other related scientific concepts, research, or theories (using the "Thinking About the Problem" section as a guide). Many analysis questions include internet searches and links to electronic resources and concept maps to promote personal reflection and further the correction of misconceptions. There are also questions for enrichment, which the teacher can use for differentiation. The expectation is that students will answer the analysis questions completely and with accuracy.
8. **Learning Targets (0 points):** This briefly lists a main objective for the lesson or concept learned while conducting the experiment.
9. **"I Learned ..." Statement (1 point):** These are facts and main ideas that apply to what students learned during the investigation. A complete sentence is expected.
10. **"Redo" Statement (1 point):** One of these sentences is due for each investigation. Students must think of a change in the variables (materials or procedures) that might result in a totally new outcome on the lab. An example might be: "I would try the experiment using a black light, rather than sunlight." A good sentence framework to use is, "Instead of using ___, I would use …".
11. **Identify One "Manipulated Variable" (1 point):** A manipulated variable is a particular component of the experiment that is purposefully changed in order to see results. Students should list the main manipulated variable, but do not need a complete sentence.
12. **Identify One "Measured Variable" (1 point):** A measured variable is the evidence of the experiment, which was observed, measured, and recorded

in a data table. Students list one of the measured variables, but do not need a complete sentence.

13. **Identify One "Controlled Variable" (1 point):** A controlled variable is held constant, and it should remain unchanged during the lab. It allows the student to determine what, if any, change took place in their variables during the experiment. Students can list one of the controlled variables, but do not need a complete sentence.

14. **Glossary (0 points):** This is a required section of any notebook, whether or not it is electronic. Definitions of terms can be used as flash card starters, as well. It serves mainly as a study guide and help for analysis and flash card generation but is not a graded portion of each lab report.

15. **Reference Pages (0 points):** These are also a required section of any notebook, whether or not it is electronic. They serve mainly as study guides and help for analysis but are not graded portions of each lab report.

References

During the development and field-testing of both editions of *Earth Science Success*, care was taken to produce a curriculum that would complement well-known Earth science print materials through a research-proven investigation methodology. Among the works consulted, three held the greatest influence: the National Science Teachers Association's four-volume series *Project Earth Science* (Ford and Smith 2000); the two-volume *Hands-on Science* series (Fried and McDonald 2000a, 2000b); and the *Curriculum Research and Development Group* series (Pottenger and Young 1992). Each of these would constitute a valuable resource for teachers who have chosen the lab-centered activities of *Earth Science Success* as their main source of lesson plans and student handouts. Along with the great ideas suggested during field-testing by colleagues, we are also indebted to the National Aeronautics and Space Administration (NASA). Two summers spent at NASA's Space Academy for Educators were instrumental in the original decision to write this book.

American Association for the Advancement of Science (AAAS). 1999. *Science for all Americans.* New York: Oxford University Press.

American Association for the Advancement of Science (AAAS). 2007. *Atlas of science literacy.* 2 vols. Washington, DC: AAAS.

Campbell, J. R., C. M. Hombo, and J. Mazzeo. 2000. NAEP 1999: *Trends in academic progress, Three decades of student performance.* Washington, DC: U.S. Government Printing Office.

Ford, B. A. 2001. *Project Earth science: Geology.* Arlington, VA: NSTA Press.

Ford, B. A., and P. S. Smith. 2000. *Project Earth science: Physical oceanography*. Arlington, VA: NSTA Press.

Fried, B., and M. McDonnell. 2000a. *Walch hands-on science series: Our solar system.* Portland, ME: J. Weston Walch.

Fried, B., and M. McDonnell. 2000b. *Walch hands-on science series: Rocks and minerals.* Portland, ME: J. Weston Walch.

Herr, N., and J. Cunningham. 2007. *The sourcebook for teaching science.* San Francisco, CA: Jossey-Bass.

National Assessment of Educational Progress (NAEP). 2012. *The nation's report card: Science 2011.* NCES Number 2012465.

National Assessment of Educational Progress (NAEP). 2013. *2011 NAEP-TIMSS linking study: Linking methodologies and their evaluations*. NCES Number 2013469.

National Research Council (NRC). 1996. *National science education standards.* Washington, DC: National Academies Press.

National Research Council (NRC). 2012. *A framework for K–12 science education: Practices, crosscutting concepts, and core ideas.* Washington, DC: National Academies Press.

National Science Teachers Association (NSTA). 2004. *NSTA position statement: Scientific inquiry. www.nsta.org/about/positions/inquiry.aspx.*

NGSS Lead States. 2013. *Next Generation Science Standards: For states, by states.* Washington, DC: National Academies Press. *www.nextgenscience.org/next-generation-science-standards.*

Pew Research Center. 2013. *Teens and technology 2013. www.pewinternet.org/~/media/Files/Reports/2013/PIP_TeensandTechnology2013.pdf.*

Pottenger, F. M., and D. B. Young. 1992. *The local environment: FAST 1, foundational approaches to science teaching.* Honolulu: University of Hawaii Curriculum Research and Development Group.

Smith, P. S. 2001. *Project Earth science: Astronomy.* Arlington, VA: NSTA Press.

Smith, P. S., and B. A. Ford. 2001. *Project Earth science: Meteorology.* Arlington, VA: NSTA Press.

Stigler, J. W., and J. Hiebert. 1999. *The teaching gap: Best ideas from the world's teachers for improving education in the classroom.* New York: Free Press.

About the Authors

Catherine Oates-Bockenstedt

Catherine has been a teacher of science at the middle-school level for almost 30 years and currently teaches at Central Middle School in Eden Prairie, Minnesota. She received both her BA and MA in science education from the University of Northern Iowa. Certified by the National Board for Professional Teaching Standards (NBPTS) in early adolescent science, she credits the professional development opportunities she has had with NBPTS and with NASA Space Academy for Educators for providing the impetus to write *Earth Science Success*. Wife of Paul, a natural resources biologist, and mother of Lara and Daniel, she lives with her family in Eden Prairie, Minnesota. She is very grateful for the support of her family, friends, and colleagues.

Michael D. Oates

Without the contributions of Michael D. Oates, this book never would have been a possibility. Michael was a professor at both the secondary and university levels for 42 years. Michael received his PhD in French and Linguistics from Georgetown University and is the author of two French textbooks. In 2008, he was decorated by the French government as *Chevalier dans l'ordre des Palmes Académiques*. Invited by his daughter, Catherine, to collaborate on *Earth Science Success*, he became an avid student of science. He was always very grateful for the support of his wife (Catherine's mother), Maureen. Diagnosed with an aggressive form of brain cancer, shortly after seeing *Earth Science Success* become published by the National Science Teachers Association, he passed away in April of 2009.

About These Labs

Each of the labs in every chapter of this book is organized to follow a pattern of active involvement by students. Students are continually presented with searching for evidence using a three-step discovery approach. The three steps are: anticipation, evidence collection, and analysis. *Anticipation* involves reflection on observations and a problem statement, recall of previous knowledge about the topic, discussion of misconceptions, and definition of concepts. *Evidence collection* includes hands-on laboratory investigation techniques. *Analysis* requires confirmation or rejection of results, reporting the findings, and making conclusions about the observations.

The hope is that students will form good habits about testing and controlling all possible variables in their experiments whenever they are collecting evidence. They should be able to identify the manipulated, measured, and controlled variables in each experiment. Results should be reliable and valid. And students should set up controls, as a basis of comparison, so they can determine the actual changes in their data. This pattern of active involvement by students is followed throughout *Earth Science Success.*

Please see the sections found in our introduction for more specifics on successful approaches for each of the labs and lessons, especially "To the Earth Science Teacher" and "Expectations for Each Investigation." In addition, teacher notes are provided to clarify differentiation possibilities, and answers are given whenever the lesson requires one particular response.

1
PROCESS OF SCIENCE
AND ENGINEERING DESIGN

1A. Testing Your Horoscope Lab

PART 1

Directions

In your lab group, briefly discuss and develop a plan for how you could test whether a horoscope predicts accurately how your day will go, discovering whether the statement can be considered reliable and valid. In outline form, write your experimental plan. Prepare to explain your design to the class. When other groups present their ideas, take note of any ideas you have for how their plans could have been improved.

Use the internet to find an example horoscope from an online newspaper. Think about which statements could be tested in an experiment. Write a potential experimental plan below:

EXPERIMENTAL PLAN FOR TESTING YOUR HOROSCOPE LAB

1. Problem:
2. Manipulated variable:
3. Measured variables:
4. Hypothesis (either "if/then" or "because"):
5. Two controlled variables:
6. Control group:
7. Number of trials:
8. Specific evidence you will collect in your data table:

PART 2

Teacher Note: See Figure 1.1 for a student sample.

Directions

In your lab group, choose one new investigation statement from the list below. Write that statement at the top of your next blank page. As a lab group, design an experiment to test the validity of that claim. Explain your experimental plan in bullet form, as you did for your horoscope, using the same format. You will be sharing the design of this experiment with the class, in a presentation.

Potential Statements for Experimentation

- Eating chocolate causes zits.
- Shaving makes hair grow back more densely.
- Drinking coffee will stunt a child's growth.
- If you swim immediately after eating, you will get cramps.
- If you go outside when your head is wet, you'll catch a cold.
- Feed a cold; starve a fever.
- Break a mirror and you will have seven years of bad luck.
- If you blow all your candles out on the first puff, you will get your wish.
- The full moon makes people restless.
- Eating carrots improves eyesight.
- If you cross your eyes too often, they will stay that way.
- Reading in dim light damages a person's eyes.
- The bread will always fall buttered side down.
- You cannot pay clear attention and text at the same time.
- An apple a day keeps the doctor away.
- Your body temperature indicates what mood you are in.

Teacher Note: A wonderful engineering design application for this final statement is found in the Arduino Starter Kit (*http://arduino.cc/en/Main/ArduinoStarterKit*), which provides students with common electronic components and engineering project ideas.

FIGURE 1.1.

SAMPLE FROM STUDENT'S ELECTRONIC SCIENCE NOTEBOOK

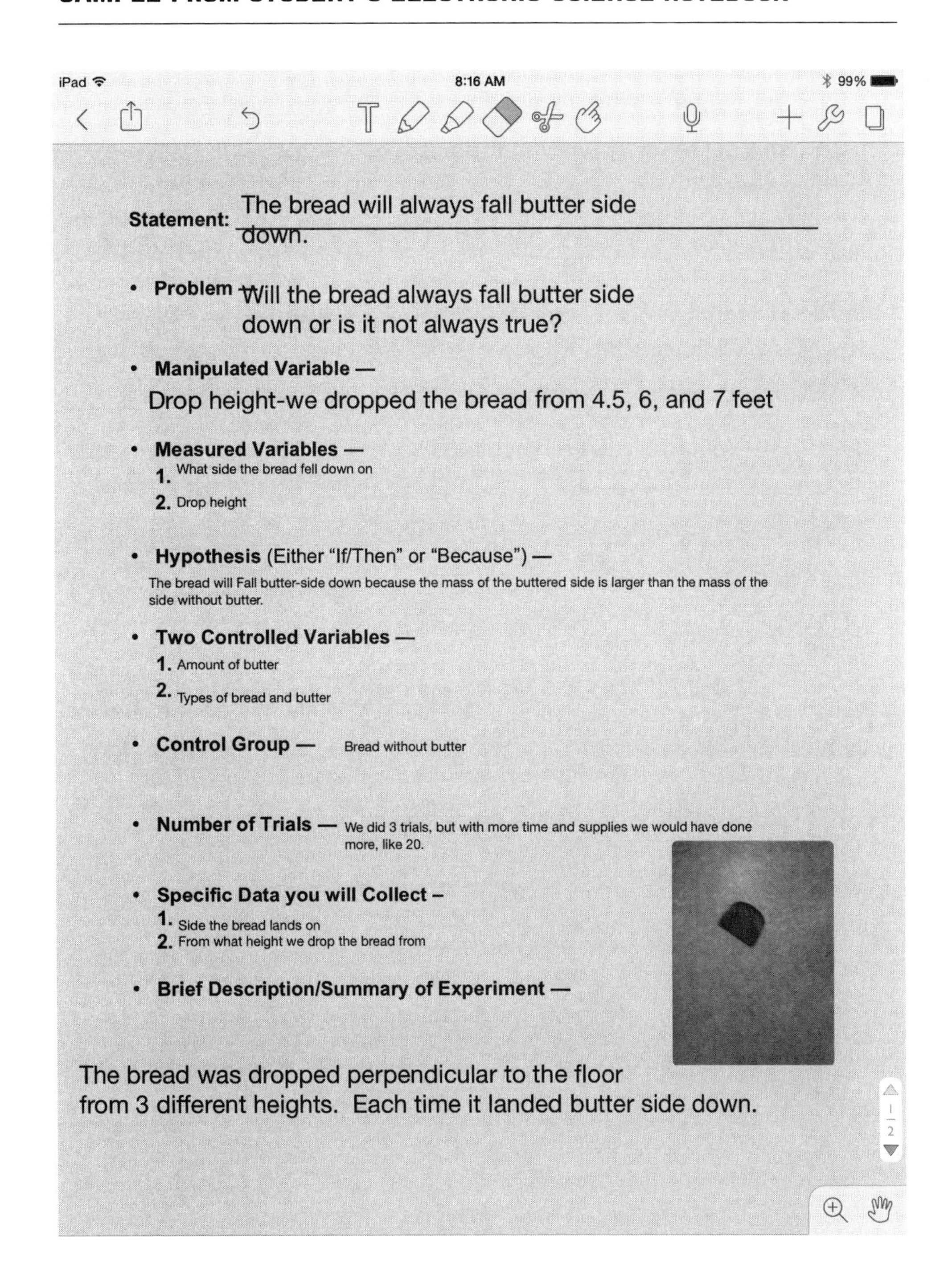

1A. Testing Your Horoscope Lab

NGSS Alignment

MS-ETS1-1. Define the criteria and constraints of a design problem with sufficient precision to ensure a successful solution, taking into account relevant scientific principles and potential impacts on people and the natural environment that may limit possible solutions.

MS-ETS1-3. Analyze data from tests to determine similarities and differences among several design solutions to identify the best characteristics of each that can be combined into a new solution to better meet the criteria for success.

3-5-ETS1-1. Define a simple design problem reflecting a need or a want that includes specified criteria for success and constraints on materials, time, or cost.

3-5-ETS1-2. Generate and compare multiple possible solutions to a problem based on how well each is likely to meet the criteria and constraints of the problem.

3-5-ETS1-3. Plan and carry out fair tests in which variables are controlled and failure points are considered to identify aspects of a model or prototype that can be improved.

1B.

Reading Minds Lab

Problem

Is the ability to read minds an example of pseudoscience?

Prediction

Explain why you think you do, or do not, have the ability to read minds.

(*Teacher note:* Have students share several predictions out loud, so misconceptions can be anticipated and explained.)

Materials

- 10 triangle cards
- 10 square cards
- 10 circle cards
- 10 star cards

Procedure

1. Cut out the squares (Figure 1.2, p. 12) to make your deck of cards for this Reading Minds Lab.
2. Work in groups of three, using the following roles: Subject, Tester, and Recorder. Rotate the roles after each complete test to assure everyone gets a turn at each role.
3. The Tester and Subject will sit across from each other, with the Recorder on the side.
4. The Tester will look at each card, think about the shape, and then hold it face down, asking the subject to read his/her mind enough to determine the shape.
5. The Recorder should make sure that the subject does not take more than three seconds to make the prediction. In other words, spontaneity is required.

6. After each prediction, the card should be passed to the Recorder who will record the Subject's prediction as well as the actual card symbol (using the Subject's data table [Data Table 1.1, p. 9]). The Subject should not be told if the prediction was right or wrong, to prevent counting of shapes.
7. Rotate roles so that someone else is the new Subject to be tested, and everyone gets a chance at each role.
8. Include a labeled sketch in the box below, showing the arrangement or lab setup you'll have for this investigation.

DRAW YOUR LABELED SKETCH HERE.

DATA TABLE 1.1.

STUDENT PROVIDES THE DESCRIPTIVE TITLE FOR THIS DATA TABLE

TRIAL #	PREDICTION	OBSERVATION	TRIAL #	PREDICTION	OBSERVATION
1			21		
2			22		
3			23		
4			24		
5			25		
6			26		
7			27		
8			28		
9			29		
10			30		
11			31		
12			32		
13			33		
14			34		
15			35		
16			36		
17			37		
18			38		
19			39		
20			40		

Analysis

1. What was your total number of correct predictions?
2. Calculate the percentage of your predictions that were correct.
3. Judging from the number of cards that you correctly predicted, do you feel that you actually do have the ability to read minds? Why? (Refer to your original prediction in answering this question.)
4. What do you think is the minimum number of correct predictions needed to be able to claim that you can read minds? You should back up your opinion with evidence, supporting your conclusion.
5. Name at least two things that might have interfered with this lab, making it unreliable or invalid for a test of reading minds.
6. What are at least two things that we were careful to keep constant in this lab? In other words, name two controlled variables.
7. Do you think 40 trials were enough, too many, or too few to yield reliable results? Explain.
8. Do you think there really is such a thing as reading minds? Explain.
9. In the Reading Minds Lab, you did not get to see the cards as you were guessing. If you were allowed to see each card after you guessed, would you be able to predict the next card more accurately? Explain why or why not.
10. (Enrichment) First, it's always good to be a bit skeptical, especially when it comes to pseudoscience. Pseudoscience is, literally, false science. Pseudoscience follows the patterns, but not the methods, theories, or systems of science. Examples of pseudoscience, which are considered to have no scientific basis, are astrology, telekinesis, and clairvoyance. Why is "mind reading" considered pseudoscience?
11. (Enrichment) How do some of the habits of mind, or attitudes in thinking, such as skepticism and curiosity, help you to better understand this activity?
12. (Enrichment) Calculate the probability of each of the following scenarios. Write your answer in both fraction and percentage form. Unlike the lab procedure, assume that each chosen card is returned to the deck, which is then shuffled before the next card is selected. Show your math for all. (*Teacher Note:* Answers are in parentheses.)

a. What is the probability of drawing a square? (¼ or 25%)

b. What is the probability of drawing squares three times in a row? (1/64 or 1.56%)

c. What is the probability of drawing the same shape two times in a row? (1/16 or 6.25%)

d. What is the probability of drawing three cards with none of them being a circle? (27/64 or 42%)

e. What is the probability of drawing all four shapes in only 4 draws? (24/256 or 0.09%)

f. What is the probability of drawing all four shapes in alphabetical order? (1/256 or 0.004%)

g. Imagine someone has guessed the correct shape eight times in a row. What is the probability that his/her next guess will also be correct? (¼ or 25%)

Learning Target

Evaluate evidence to determine whether a claim is valid.

I Learned:

Redo:

Manipulated Variable:

Measured Variable:

Controlled Variable:

FIGURE 1.2.

DECK OF CARDS FOR THE READING MINDS LAB

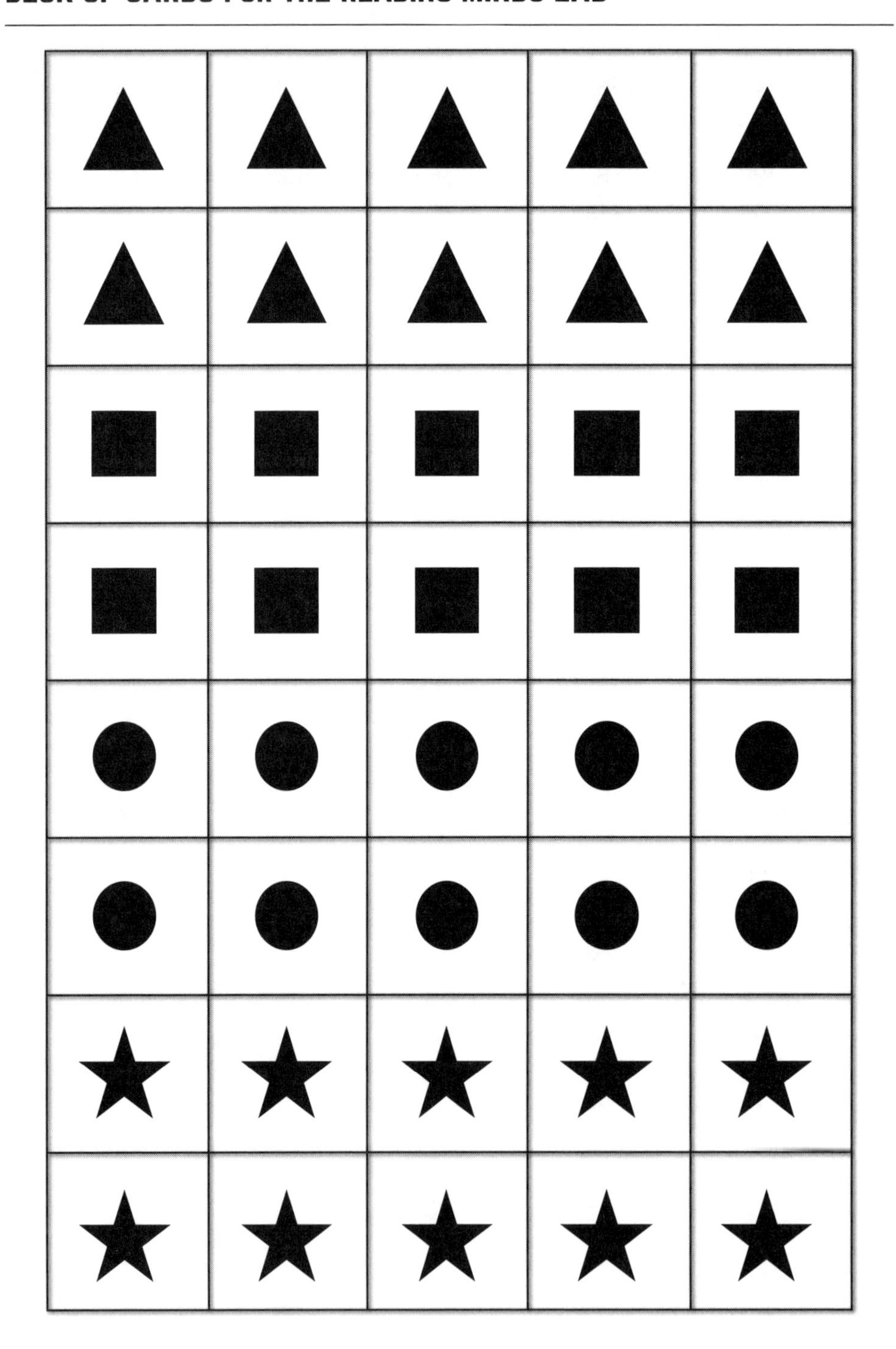

1B. Reading Minds Lab

NGSS Alignment

MS-ETS1-1. Define the criteria and constraints of a design problem with sufficient precision to ensure a successful solution, taking into account relevant scientific principles and potential impacts on people and the natural environment that may limit possible solutions.

MS-ETS1-3. Analyze data from tests to determine similarities and differences among several design solutions to identify the best characteristics of each that can be combined into a new solution to better meet the criteria for success.

3-5-ETS1-1. Define a simple design problem reflecting a need or a want that includes specified criteria for success and constraints on materials, time, or cost.

3-5-ETS1-2. Generate and compare multiple possible solutions to a problem based on how well each is likely to meet the criteria and constraints of the problem.

3-5-ETS1-3. Plan and carry out fair tests in which variables are controlled and failure points are considered to identify aspects of a model or prototype that can be improved.

1C.

Estimating With Metrics Lab and Measurement Formative Assessment

Problem

Which equipment is best to use when measuring liquids and solids?

Prediction

Describe, in one sentence, how you would measure a solid differently from a liquid.

(*Teacher Note:* Have students share several predictions out loud, so misconceptions can be anticipated and explained.)

Thinking About the Problem

Can you name any advantages of the metric system? Does it make any difference if we measure in yards and feet versus meters and centimeters? Scientific collaboration is one advantage of the metric system. Today scientists collaborate with colleagues all over the world and so it is vital that they have similar tools and standards of measurement at their disposal.

Two of the most important process skills that scientists use every day are estimations of measurement and manipulation of standard laboratory equipment. Without skill in these two areas, accurate observations would be impossible. Estimating prior to measuring helps us check for accuracy in our measurements. This is especially true when it comes to modeling the vast distances found in the study of space science.

Background knowledge of the metric system lends a globally understood language to findings by scientists. Many scientists from across the world are responsible for discoveries in astronomy, and need to communicate effectively for the findings to be widely understood.

In this lab, you will enhance your skills in the areas of metric conversions, estimations of measurement, and the manipulation of standard laboratory equipment.

Write three main points from the "Thinking About the Problem" reading:

1.

2.

3.

Procedure

1. First estimate, and then use the triple beam balance to measure 15 miscellaneous objects from around the room in grams (Data Table 1.2, p. 17).
2. First estimate the length of the following, and then use a metric ruler to find the exact length in centimeters: your arm, the room, a textbook, your height, and a desk (Data Table 1.3, p. 18).
3. First estimate the temperature of the following, and then use a metric thermometer to find the exact temperature in degrees Celsius: room temperature, hot tap water, and cold tap water (Data Table 1.4, p. 18).
4. First estimate the volume of water in the bucket provided by your teacher, and then use a graduated cylinder to determine the exact volume in milliliters (Data Table 1.5, p. 19).

Learning Target

Enhance skills in the areas estimations of metric measurement and the manipulation of standard laboratory equipment.

I Learned:

Redo:

Manipulated Variable:

Measured Variable:

Controlled Variable:

Analysis

1. Which equipment should be used to measure liquids? Why?
2. Which equipment should be used to measure solids? Why?
3. Draw your labeled sketch of your lab equipment in the box below.

DRAW YOUR LABELED SKETCH HERE.

DATA TABLE 1.2.

PROCEDURE #1 ON ESTIMATING MASS

OBJECT	ESTIMATED MASS	ACTUAL MASS
1		
2		
3		
4		
5		
6		
7		
8		
9		
10		
11		
12		
13		
14		
15		

DATA TABLE 1.3.

PROCEDURE #2 ON ESTIMATING DIMENSIONS

OBJECT	ESTIMATE	ACTUAL
Your right arm (shoulder to finger tip)		
The classroom (North wall to South wall)		
A textbook (top to bottom edge)		
Your height (head to toe)		
A desk (surface to floor)		

DATA TABLE 1.4.

PROCEDURE #3 ON ESTIMATING TEMPERATURES

SUBJECT	ESTIMATE	ACTUAL
Room temperature (°C)		
Hot tap water (°C)		
Cold tap water (°C)		

DATA TABLE 1.5.

PROCEDURE #4 ON ESTIMATING VOLUME

CONTAINER	ESTIMATE	ACTUAL
Bucket of water (ml)		

4. Circle the most sensible measurement: (*Teacher Note:* Answers are given in parentheses.)

Length of football field (C):	A. 90 mm	B. 90 cm	C. 90 m	D. 90 km
Height of a woman (B):	A. 160 mm	B. 160 cm	C. 160 m	D. 160 km
Width of a room (C):	A. 8 mm	B. 8 cm	C. 8 m	D. 8 km
Width of a desk (B):	A. 75 mm	B. 75 cm	C. 75 m	D. 75 km
Mass of dog (C):	A. 8 mg	B. 8 g	C. 8 kg	D. 8 metric tons
Mass of pencil (B):	A. 5 mg	B. 5 g	C. 5 kg	D. 5 metric tons
Mass of raisin (B):	A. 1 mg	B. 1 g	C. 1 kg	D. 1 metric ton
Mass of human (C):	A. 60 mg	B. 60 g	C. 60 kg	D. 60 metric tons
Volume of car gas tank (B):	A. 8 ml	B. 8 L	C. 80 ml	D. 80 L
Volume of coffee cup (A):	A. 250 ml	B. 250 L	C. 25 ml	D. 25 L
Volume of bathtub (D):	A. 400 ml	B. 400 L	C. 40 ml	D. 40 L
Volume of pop can (A):	A. 500 ml	B. 500 L	C. 5 ml	D. 5 L
Temp of boiling water (A):	A. 100°C	B. 50°C	C. 212°C	D. 1000°C
Temp of ice (A):	A. 0°C	B. 32°C	C. 60°C	D. 100°C
Room temperature (A):	A. 22°C	B. 55°C	C. 72°C	D. 100°C
Body temperature (B):	A. 97°C	B. 37°C	C. 310°C	D. 100°C

Sample Formative Assessment on Measurement

Directions

You will be taking this formative assessment on your iPads. Do your best, on your own. My goal, as your teacher, is to determine whether you understand how to read measurements from a triple beam balance, a graduated cylinder, a ruler, and a thermometer. Do your best to determine exact measurements, which include two decimal places and units. (*Teacher note:* To see this document with the assessment questions, students may view it from your Google website and then respond to the Google form, or they can respond to the questions using a document uploaded to *Schoology*. With iPads, students are able to four-finger-swipe between two documents, if needed.)

Assessment Questions

1. Look carefully at the image below. Accurately determine the exact measurement shown on the triple beam balance. Record your answer on the Google Form (Figure 1.3).

(*Teacher note:* Answer = 272.52 g)

2. Look carefully at the image below. Accurately determine the exact measurement shown on the graduated cylinder. Record your answer on the Google Form (Figure 1.4).

(*Teacher note:* Answer = 11.50 ml)

3. Look carefully at the image below. Accurately determine the exact measurement shown on the thermometer. Record your answer on the Google Form (Figure 1.5).

(*Teacher note:* Answer = 31°C)

FIGURE 1.3

TRIPLE BEAM BALANCE

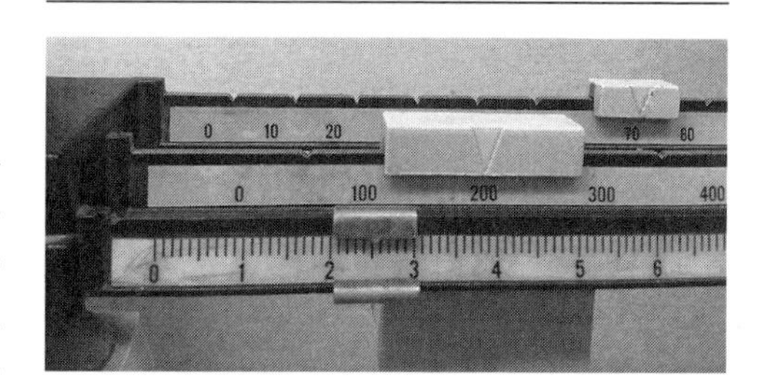

FIGURE 1.4.

GRADUATED CYLINDER

FIGURE 1.5.

THERMOMETER

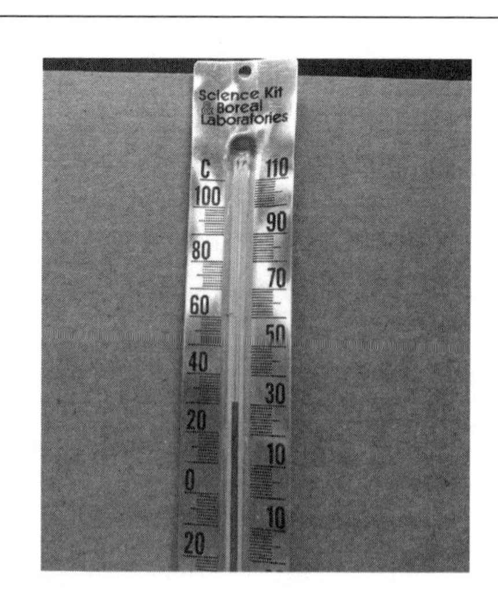

4. Look carefully at the image below. Accurately determine the exact length of the rectangle shown above the ruler. Record your answer on the Google Form (Figure 1.6).

(*Teacher note:* Answer = 8.35 cm)

FIGURE 1.6.

RULER

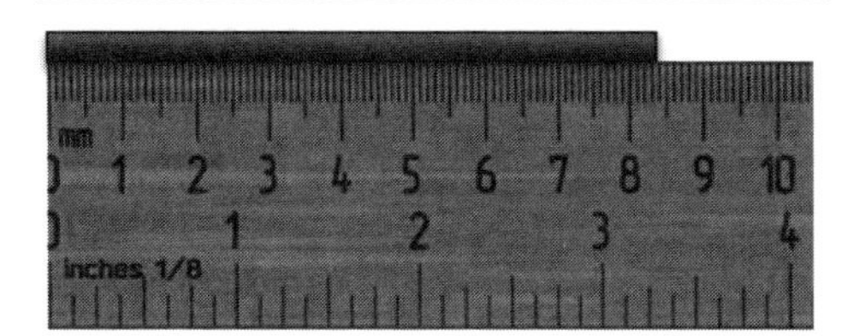

1C. Estimating With Metrics Lab and Measurement Formative Assessment

***NGSS* Alignment**

MS-ETS1-1. Define the criteria and constraints of a design problem with sufficient precision to ensure a successful solution, taking into account relevant scientific principles and potential impacts on people and the natural environment that may limit possible solutions.

3-5-ETS1-3. Plan and carry out fair tests in which variables are controlled and failure points are considered to identify aspects of a model or prototype that can be improved.

1D.

Science Process Vocabulary Background Reading and Panel of Five

Rules and Procedures for Panel of Five

1. Use this reading/review game for the five background readings in this book. Panel of Five is a fun strategy used to teach students about the concepts in any science-related newspaper or magazine article. The teacher needs to make about 35 paper copies of the article, and have six inexpensive prizes available for awarding per class.
2. Panel of Five begins by randomly selecting one student to be "director" and five students to start out as "panel members." The rest of the students will sit in a half-circle facing the five panel members. The diagram below (Figure 1.7) shows the classroom arrangement:

FIGURE 1.7.

CLASSROOM ARRANGEMENT FOR PANEL OF FIVE

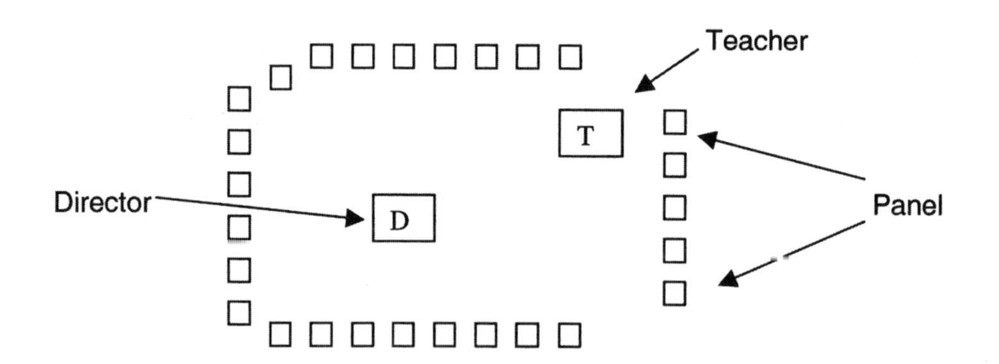

3. There are three rules in Panel of Five. First, no fill-in-the-blank style questions can be asked. Students must instead ask good, challenging questions that start with a what, where, when, why, how, or even true/false. Second, once a panel member has answered a question correctly, he or she does not need to answer another question until the next paragraph is read. Third, one spelling question is allowed per paragraph. This sometimes helps to temporarily remove a hard-to-eliminate student from the panel, especially when he or she does not get to see the words.
4. The panel members do not get to have an article in front of them (they must use only listening skills). The teacher's job (along with classroom management) is to read clearly, paragraph by paragraph, whenever the director designates. The student director runs the classroom, calling on four or five students per paragraph (turn by turn). The student, when called upon, would ask a particular panel member a question about the paragraph that was just read. If the panel member answers it correctly, then the panel member gets to stay on the panel. If the panel member answers it incorrectly, then the asking student switches places and gets to be on the panel. The director says "switch" or "stay" after the panel member has given his or her response, and then the director calls on the next student to ask a question.
5. At the end of class, the five final panel members each get a prize, and so does the director. Students can cover about 8–10 paragraphs in an average 50-minute class period.

Science Process Vocabulary Background Reading (for Use in Panel of Five)

Authentic science involves finding the answers to questions. There are two ways to find the answers to questions about which students are curious. One is to research the answer extensively, and write a report about your findings. The other involves conducting a well-controlled experiment, which includes hypothesizing, testing through experiments, controlling variables, collecting evidence, doing background research, analyzing data, and reporting results. Often, experiments can be used to determine the effects of one variable on another variable. For example, an investigation could determine the effects of various natural materials on water purity.

One should begin by explicitly stating the experimental question being investigated. This is referred to as the problem statement. The question should provide enough detail so the audience can understand what will be done in the experiment.

Phrase the issue you are trying to resolve as a question. For example: How does the color of test paper affect student performance on a math test?

The hypothesis is an explanation based on observations, which can be tested by experiment. It is not a prediction (an educated guess), but it might be based on a prediction. Sometimes it helps to think of the hypothesis as an "if/then" prediction or a prediction with a testable "because…." The hypothesis answers how changing the manipulated variable will affect the measured variables. Examples include: If I increase the engine size, then the rocket will fly higher and O'Leary Lake will have clearer water because it has buffer zones of tall grasses on all of its shorelines.

Thinking about the background information related to the problem statement is important. A report should be made about the research related to the chosen experimental topic. This research should define and explain all of the major scientific concepts involved in the experiment, including information on what remains known and unknown about the topic.

While conducting the experiment, all materials and procedures should be documented so that the experiment is replicable by others. This lends validity to the results. Make certain to test and control all possible variables in the experiment. A variable includes any characteristics in an experiment that change, or could be changed. Identify the *manipulated variable* and *measured variable*. The manipulated variable is what is being tested, or purposefully changed (x-axis on most graphs). It is the variable changed on purpose, in order to see its effect. The measured variable is the result being measured (y-axis on most graphs). It is the variable that responds to a change in the manipulated variable, and is also called the "responding variable."

Remember that a good experiment only tests one purposefully changed variable at a time. Hold all other variables constant, while you are testing the one variable. Set up a control as a basis of comparison, so you can determine the actual changes in your experiment. Constants, or controlled variables, include characteristics in an experiment that are kept unchanged in all trials. These variables could affect the outcome of the experiment, so they are kept the same (held constant) each time the experiment is carried out. The control is the standard for comparison in an experiment. Repeat your trials and collect adequate data in order to give reliability, as well. Trials are the number of times an experiment is repeated for each level or value of the manipulated variable. The more trials conducted, the more reliable the results will be.

The results should contain data tables, graphs, images, charts, tables, and day-by-day logs. All graphs, tables, images, and charts should be descriptively labeled so their content can be fully understood. Good scientists keep a journal of progress and thoughts during the experiments, as well. In the results section, a discussion

about the types and quantities of evidence collected is important. Discuss exactly what was observed and measured. Looking over the data tables, graphs, charts, or daily log, and then explaining what the evidence shows or seems to indicate, is vital.

In the conclusion, compare the original hypothesis with the results. Describe potential sources of error. Describe how the results are supported by other related scientific concepts, research, or theories (background research is helpful with this). It is also helpful to describe how this particular experiment should be done differently, if it were to be done again.

1D. Science Process Vocabulary Background Reading and Panel of Five

NGSS Alignment

MS-ETS1-1. Define the criteria and constraints of a design problem with sufficient precision to ensure a successful solution, taking into account relevant scientific principles and potential impacts on people and the natural environment that may limit possible solutions.

3-5-ETS1-3. Plan and carry out fair tests in which variables are controlled and failure points are considered to identify aspects of a model or prototype that can be improved.

1E.

Explain Everything With Science Trivia

Scientists are basically problem-solvers who work to find the best possible answers to questions about nature. The questions that we have been working on in labs are answered by experimentation. Some questions in science, however, are answered by research. You will use the *Explain Everything* app on your iPad to teach your audience the answer to five science research questions. First, use appropriate websites to find answers to all of the following questions. Then, create an *Explain Everything* lesson, which teaches any five of the concepts in a memorable way.

Work on your own. Select any five of the trivia questions below. Some of these trivia questions are quite easy. Some of them will correct a misconception that you had about the facts. And some of them will be quite challenging. (*Teacher Note:* Answers are in parentheses.)

1. Name the eight classical planets in order of increasing distance from the Sun. (Mercury, Venus, Earth, Mars, Jupiter, Saturn, Uranus, Neptune)
2. What is the highest mountain in the world, above sea level? (Everest) What is the highest mountain that starts below sea level? (Mauna Kea)
3. What pigment makes leaves green? (Chlorophyll)
4. Which light, blue or red, has higher energy? (Blue)
5. What are the three main parts of an atom? (Electron, proton, neutron)
6. Would you be going further if you went 10 miles or 10 kilometers? (10 miles)
7. What is a meteorologist? (Scientist who forecasts weather conditions)
8. What is the common name, in the United States, for the star Polaris? (North Star)
9. What does a botanist study? (Plants)
10. What are the chemical symbol and the atomic number for iron? (Fe 26)

11. What is created when an extremely large star collapses at the end of its life cycle? (Black hole)
12. What is the largest living species of feline? (Siberian tiger)
13. During which era did dinosaurs rule the land? (Mesozoic)
14. What two months contain an equinox? (March and September)
15. What are the three main fossil fuels? (Coal, oil, natural gas)
16. What galaxy are we in? (Milky Way)
17. What constellation represents a hunter with a sword and belt? (Orion)
18. What virus is believed to cause AIDS? (Human Immunodeficiency Virus)
19. Of the following, which was the largest dinosaur: Tyrannosaurus Rex, Seismosaurus, or Apatosaurus? (Seismosaurus)
20. Which comet can we see approximately every 76 years? (Halley)
21. What bird lays the largest egg? (Ostrich)
22. What does a sling psychrometer help scientists measure? (Relative Humidity)
23. Which of the following machines did Johannes Gutenberg invent: ballpoint pen, printing press, or cotton gin? (Printing press)
24. What do isobars represent? (Lines of equal atmospheric pressure)
25. What is the only day of the week named for a planet? (Saturday)
26. What are the four major blood groups? (A, B, AB, O)
27. How many chambers are there in the human heart? (4) How many chambers does a frog's heart have? (3)
28. What does DNA stand for? (Deoxyribonucleic Acid)Where is DNA found? (In cells)
29. If you are increasing in latitude in the Northern Hemisphere, are you traveling east, west, north, or south? (North)
30. What is the common name, in the United States, for the constellation found within Ursa Major? (Big Dipper)
31. Is a spider an insect? (No, arachnid)

32. Every day, an average of 190 liters of blood is cleaned in your kidneys, lungs, or liver? (Kidneys)
33. What turns blue litmus paper pink? (Acid)
34. What two things do bees collect? (Pollen, nectar)
35. What part of the body does glaucoma strike? (Eye)
36. On whom would you use the Heimlich maneuver? (Choking victim)
37. What scale measures earthquakes? What machine generates this measurement? (Moment Magnitude Scale, Seismograph) (Teacher Note: The MMS was developed in the 1970s to replace the Richter scale.)
38. If you have a condition called "bromhidrosis," do you have smelly breath, perfume, or armpits? (Armpits)
39. How did pterodactyls get from one place to another? (Gliding)
40. What are the three major groups of rocks? (Igneous, metamorphic, sedimentary)
41. What is the name of the point at which condensation begins? (Dew point)
42. Which two months contain a solstice? (June and December)
43. What causes the Earth to have seasons? (Tilt of axis as Earth travels around Sun)
44. Would you find your philtrum under your nose, tongue, or chin? (Nose)
45. For what work in physics did Einstein receive his Nobel Prize? (Photoelectric effect)
46. What does RADAR stand for? (Radio Detecting and Ranging)
47. What does LASER stand for? (Light Amplification by Stimulated Emission of Radiation)
48. Name the eight groups used in the Classification of Living Things, starting with Domain. (Domain, Kingdom, Phylum, Class, Order, Family, Genus, Species)
49. Why does Mars get more ultraviolet radiation from the Sun than the Earth does? (Less Atmosphere)

1E. Explain Everything With Science Trivia

NGSS Alignment

MS-ETS1-1. Define the criteria and constraints of a design problem with sufficient precision to ensure a successful solution, taking into account relevant scientific principles and potential impacts on people and the natural environment that may limit possible solutions.

MS-ESS1-1. Develop and use a model of the Earth-Sun-Moon system to describe the cyclic patterns of lunar phases, eclipses of the Sun and Moon, and seasons.

3-5-ETS1-3. Plan and carry out fair tests in which variables are controlled and failure points are considered to identify aspects of a model or prototype that can be improved.

1F.

Controlled Experiment Project

Sample Letter to Students and Parents

This letter describes the requirements for the Controlled Science Experiment Project. Authentic science involves finding the answers to questions. There are two ways to find the answers to questions about which students are curious. One (which will not be our main focus at this time) is to research the answer and write a report. The other (which is a controlled experiment, and will be the focus of this assignment) involves hypothesizing, testing through experiments, controlling variables, collecting evidence, doing background research, analyzing data, and reporting results. Students must develop a controlled experiment (not a research paper) in one of three science fields: Earth science, life science, or physical science, using the science inquiry process. Students have been given quite a bit of information regarding the process of science in class.

The Experiment Project Category Options (You must pick one.)

Earth science includes (but is not limited to) environmental issues, astronomy, meteorology, geology, weathering/erosion, topography, water quality, and soil studies.

Physical science includes (but is not limited to) chemistry, forces, acceleration, renewable energy sources, magnetism, light and optics, acid and base reactions, and electricity.

Life sciences includes (but is not limited to) anatomy, health/wellness, human behavior, plant growth, and pesticide/fertilizer/herbicide applications. The testing of animals with fur, feathers, or fins is not allowed.

(*Teacher note:* Categories can be modified to include consumer science, as well.)

Students will not be allowed to work in groups for this assignment. A keynote presentation, using student iPads, is required. The students will be delivering this presentation to the class. Parents are invited to attend their child's presentation.

This entire project is made up of small assignments, culminating in a class presentation, using the *Keynote* app on our iPads. This project can be thought of as a

do-it-yourself, real-world, laboratory experiment. The critical part is that it is an experiment, not a demonstration of how things work, and not a research paper.

Science Experiment Presentation Requirements

Following the guidelines below, create a keynote presentation on your iPad (remember to take screen shots of your slides, or back up everything to Dropbox or Google Drive). All portions must use correct grammar, punctuation, spelling, and mechanics. Have an adult proofread your text. All text should be word-processed.

Title and Category

The title of your experiment project should be several descriptive words, not a complete sentence, dealing with the chosen topic of your experiment. It should be brief but it should also indicate the variable(s) that were tested. (Poor Example: "The Best Water Purity"; A Better Example: "Investigation of the Effects of Various Natural Materials on Water Purity") The category should be one of the following sciences: Earth, life, or physical.

Problem

Explicitly state the experimental question being investigated. The question should provide enough detail so the audience can understand what will be done in the experiment. (Example: "How does the color of test paper affect student performance on a math test?")

Hypothesis

This should be a complete sentence, giving your hypothesis (an explanation based on your observations, which can be tested by experiment). It is not a prediction (an educated guess), but it might be based on a prediction. Sometimes it helps to think of the hypothesis as an "if/then" prediction or a prediction with a testable "because ..." (Examples: "If I increase the engine size, then the rocket will fly higher" or "O'Leary Lake will have clearer water because it has buffer zones of tall grasses on all of its shorelines.")

Thinking About the Problem

This requires one full page (word-processed, double-spaced, in font size 12, with one-inch margins). You must show significant background research information on your chosen experimental topic. This is not about your opinion, and it is not about

your procedure. It is about reporting the research related to your topic. Define and explain all of the major scientific concepts involved in your experiment. What is known about the topic? What remains unknown about the topic? Define and explain the major scientific concepts involved in your experiment. If applicable, state differences among the variables you will be testing. How does your experiment fit into this context? Cite your references in a bibliography.

Pointers for Good Online Background Research

For many, the first step in an online research journey is a simple Safari, Google, or Bing search. Here are some more specific ideas for your background research:

1. Start with the most-used search engines on the web (Safari, Google, or Bing). Put your phrase in quotes to return pages with the exact words, like this: "Growth of Marigolds."
2. Search with the main goal of getting an overview of the topic. Good resources are often websites ending in *.edu*, *.org*, or *.gov*. Google Scholar will give you scholarly journal articles and other verified sources of information, as well.
3. YouTube is good for videos that can teach you things and answer questions. You may be able to learn useful background information from these.
4. Khan Academy has more than 2,000 videos covering physics, biology, chemistry, and statistics. The videos are short, simple, and entertaining.
5. Follow your curiosity, but keep track of the links you're following in a Word document, or an application like *Notability*, so you can consult them later.

Materials

Create a complete list, written in a column format, of all of the materials that you use. Begin your experiment now, too!

Procedure Steps and Labeled Image

Take one photo, showing all of your assembled materials, and label the image, using *Explain Everything* or *Notability*. This labeled image should help you in clearly describing the procedure steps. The steps (maximum of five listed and numbered steps) and labeled image should enable you to explain what anyone else should do in order to replicate your same experiment. During your presentation, you will go into detail, describing your experimental procedure.

Conducting the Experiment

Make certain that you test and control all possible variables in your experiment. Identify the *manipulated variable* and *measured variable* in your experiment. The manipulated variable is what you are testing or what you purposefully change (x-axis on most graphs). The measured variable is the result you are measuring (y-axis on most graphs). For example, in an experiment measuring root length of plants vs. amount of water, root length is the measured variable because that is what you are measuring. The other variable would be the amount of water given to the plants. Remember that a good experiment only tests one purposefully changed variable at a time! Hold all other variables constant while you are testing that one variable. Set up a control as a basis of comparison, so you can determine the actual changes in your experiment. Repeat your trials and collect adequate data to give reliability, too!

Abstract

The abstract is a one-paragraph detailed summary of your findings and results. It condenses the entire experiment report into a brief, clear, quickly readable overview. Try to answer the following four questions: What were you looking for? How did you look for it? What did you find? What does this mean? The abstract should also serve to help you in completing your Community Connection.

Proof of Community Connection

Submit or show evidence/proof of your Community Connection assignment completion. To fulfill the requirements of this assignment, you will use your abstract information to do one of the following: you must publish a brief letter to the editor in our local newspaper or attend and present at a community meeting (e.g., school board, city hall meetings, church event) related to your problem statement. Your hope is to help educate the public about your issue, inform decision makers about wise decisions they could make, or advocate for new choices being made at a local community level. Look closely at your problem statement for societal connections, and then find a way to act upon what you have learned. You must have your parents proofread your information before submitting it, so that they can approve your format and statements. You must be able to submit proof of your actions by the due date. If you submit a letter to the editor, it must begin as follows:

"Dear Editor,

I am a 7th grader at ScienceRocks Middle School. For an assignment, I designed and conducted a controlled science experiment. The problem statement that I was

investigating was ___. My results should be of interest because ___." (Use some of your abstract paragraph in order to describe your experiment.) Thank them for reading at the end of your brief letter.

Results

Several photos of your results must be included. In addition, you may bring in the actual results or props on presentation day, if they help explain your procedure or results. (Props often help you demonstrate experimental materials being used.) The results section of your presentation should contain at least one data table, and may also include graphs, charts, tables, or day-by-day logs (these can be images of handwriting or drawn electronically, if necessary). Make sure that you label your graphs or charts so the audience can understand them. Photos are preferred over actual results when the experiment involves chemicals, hazardous materials, or plants. You should keep a journal of your progress and thoughts as you go through the experiment, as well.

Presentation

You will do a presentation, using the *Keynote* app your iPad. You must describe the purpose and procedure for your experiments. Discuss how your variables were controlled. Discuss what you learned from background research on your topic, as well. You must describe the types and quantities of data you collected. Discuss what you observed and what you measured. You must compare your original hypothesis with your results. Look over your data tables, graphs, charts, and daily log and then explain what you think the data show or seem to indicate. Include how your results are supported by other related scientific concepts, research, or theories (use your background research to help with this). You must report on your Community Connection assignment. Finally, you must give one "I learned…" statement, and one "Redo" statement about how you would do this particular experiment project differently, if you were to do it again. Areas of potential future research should be described here, as well. Your Keynote Presentation should follow the script provided in this packet. This means that when the script requires you to share your hypothesis, then it should be showing on the Keynote Presentation slide at that time for the audience to see.

Props and Photos

You must have several images, showing both the process and the results from your experiments. These images should appear throughout your Keynote Presentation. You must also have one labeled image, which accompanies your procedure steps. Props are not required, but often help explain and clarify complex procedures.

Multiple-Choice Questions for the Audience

An additional task is to create three challenging multiple-choice questions for your audience. Your peers should be able to answer these questions after viewing your presentation. Think through your science experiment project. Develop three questions about main points covered by your results or background research. The three questions, with multiple-choice options (a, b, and c), and correct answers, must be in written form, and will be presented to the audience at the end of your presentation.

Procedure Recap

1. Find a question that you are curious about (you could search any number of "science fair ideas" sites on the internet).
2. Determine into which of the science categories the question fits (Earth, life, or physical).
3. Form a hypothesis (an explanation based on your observations that can be tested by experiment).
4. Do background research (search for related experiments, background information on the concepts, and applicable science theories) and write the one-page *Thinking About the Problem* section.
5. Experiment and investigate (find a way to answer your problem statement by manipulating and controlling variables).
6. Compile all of the evidence and results (graphs, charts, tables, and day-by-day logs) and take photos of results. You may want to bring actual results for your presentation, as well.
7. Form a conclusion based on your results (answer the original question/problem statement) and use it to write the abstract.

8. Complete the Community Connection assignment (based on the real-world implications of your problem statement).
9. Write three multiple-choice questions, with answers, for the audience.
10. Prepare the *Keynote* presentation on your iPad, following the packet guidelines and practice the presentation (use the attached rubric and script for help in this).

Science Project Presentation Script

1. My name is …
2. The title of my project is …
3. My project is in the (life, Earth, or physical science) category.
4. My problem statement is …
5. The hypothesis I had was … (point to your *Thinking About the Problem* paragraphs on your presentation)
6. To summarize my background research, I found that … (point to the Materials List on your presentation. Show any props, if you brought them)
7. The materials that I used were … (point to the Procedure and Labeled Image on your presentation)
8. The steps for my procedure are as follows: Step One … (use labeled image to help explain each step)
9. The main variable that I was testing was …
10. The things that I was careful to control were …
11. The types of evidence that I collected included … (point to the data tables, graphs, images, and any actual results you brought in)
12. My results showed that … (again, refer to the data tables, graphs, images, and any actual results you brought in)
13. The overall summary and outcome of my investigation was … (point to your Abstract paragraph on your presentation)
14. Now, I will read my abstract aloud … (point to your Community Connection evidence)

15. What I did for my community connection was …
16. If I were to redo this experiment, I would …
17. I learned …
18. I most enjoyed …
19. My multiple-choice questions are …
20. Thank you for listening to my presentation. You may now applaud wildly for me.

1F. Controlled Experiment Project

NGSS Alignment

MS-ETS1-1. Define the criteria and constraints of a design problem with sufficient precision to ensure a successful solution, taking into account relevant scientific principles and potential impacts on people and the natural environment that may limit possible solutions.

MS-ETS1-2. Evaluate competing design solutions using a systematic process to determine how well they meet the criteria and constraints of the problem.

MS-ETS1-3. Analyze data from tests to determine similarities and differences among several design solutions to identify the best characteristics of each that can be combined into a new solution to better meet the criteria for success.

3-5-ETS1-1. Define a simple design problem reflecting a need or a want that includes specified criteria for success and constraints on materials, time, or cost.

3-5-ETS1-2. Generate and compare multiple possible solutions to a problem based on how well each is likely to meet the criteria and constraints of the problem.

3-5-ETS1-3. Plan and carry out fair tests in which variables are controlled and failure points are considered to identify aspects of a model or prototype that can be improved.

2

EARTH'S PLACE IN THE SOLAR SYSTEM AND THE UNIVERSE

2A.

Sizing up the Solar System Lab

Problem

How big are the planets?

Prediction

Describe how you think the Earth compares to the Sun in terms of size.

(*Teacher note:* Have students share several predictions out loud, so misconceptions can be anticipated and explained.)

Thinking About the Problem

What do astronomers do? The word astronomy (*astron* "star" in Greek) means literally "the study of stars" and human beings have been gazing in wonder at the sky for a very long time. Careful observations by ancient scientists are what helped reveal "tricks" or "patterns" that explain events in the night sky. These ancient stargazers were from many different cultures, including the Greeks, the Romans, the Mayans, and many African countries. They developed myths to explain events associated with things they did not understand (movements of the stars and planets, for example).

How did early astronomers determine the size of the different planets? Very careful observers from many cultures were able to explain things mathematically with reliability, predictability, and precision. It helped that there was far less light pollution long ago! They watched as planets crossed in front of the Sun, used triangulations with known angles, and applied geometry to their

FIGURE 2.1.

SOLAR SYSTEM SKETCH

(*Teacher Note:* Sketch involves labeled dots, for use in reference and learning about dwarf vs. classical planets.)

Sun: Star

Mercury: Classical Planet

Venus: Classical Planet

Earth: Classical Planet

Mars: Classical Planet

Ceres: Dwarf Planet (formerly considered an asteroid)

Jupiter: Classical Planet

Saturn: Classical Planet

Uranus: Classical Planet

Neptune: Classical Planet

Pluto: Plutoid (formerly considered a planet and a drawf planet)

Charon: Dwarf Planet (formerly considered a moon of Pluto)

Eris: Plutoid (a trans-Neptunian object in the Kuiper belt, whose discovery caused a redefining of the planets)

measurements. These observers formulated testable hypotheses and made many follow-up observations to verify their findings.

One observation that you could make is that the Earth seems very large. Yet, Earth is dwarfed by our immense Sun. Earth, with its 12,756-kilometer diameter, is still 109 times smaller than the Sun. Even Jupiter, the largest planet, is only about one-tenth the Sun's diameter. And Pluto is almost 600 times smaller than the Sun. Pluto is a good example of ongoing scientific debate over classification. Initially one of the nine planets in our solar system, it has recently been redefined twice: as a dwarf planet and then as a plutoid. This redefinition remains controversial among some scientists, and may be subject to further modification. Scientists are constantly redefining what they know based on new information.

Processes and conditions that were at work during the formation of the solar system (*sol* "sun" in Latin) about 5 billion years ago formed the planets into the sizes that we now observe. The four largest planets (Jupiter, Saturn, Uranus, and Neptune) contain enormous amounts of hydrogen and helium. Earth and the other smaller planets do not contain nearly as much of these gases. The quantities of these gases affect how large a planet can become.

In this lab, you will develop your own scale model of the planets to learn about their relative sizes, compared to the Sun.

Write three main points from the "Thinking About the Problem" reading:

1.

2.

3.

Procedure

1. Find the diameter of your model Sun in centimeters.
2. Use the model Sun's diameter to calculate the comparison constant (CC). You will then use the CC to figure out the diameters for each planet model (Data Table 2.1, p. 44). (The model Sun diameter divided by 109 equals the CC.) Once calculated, write the CC in your data table.
3. Use the data table to calculate the planet model diameters for each planet. Multiply the comparison diameter by the CC to get the diameter for each planet model. Use only two decimal places in your answer.
4. For each planet, use wadded-up scratch paper and a ruler to make a model with the required diameter.
5. Place your collection of planets near your model Sun. In the box below, use your iPad to photograph and label the planets, showing planet model sizes.

INSERT LABELED IMAGE OF YOUR MODEL HERE.

DATA TABLE 2.1.

PLANET MODEL DIAMETERS

PLANET TYPE	COMPARISON DIAMETER	COMPARISON CONSTANT	PLANET MODEL DIAMETER
Classical Planets (Large enough to be spherical and are dominant over smaller bodies in their path)			
Mercury	0.38		
Venus	0.95		
Earth	1.00		
Mars	0.53		
Jupiter	11.20		
Saturn	9.50		
Uranus	4.00		
Neptune	3.90		
Dwarf Planets (Large-diameter objects, but unable to clear smaller bodies in their path)			
Ceres	0.08		
Charon	0.10		
Plutoids (Dwarf planets that are Pluto-sized or larger)			
Pluto	0.19		
Eris	0.29		

Analysis

1. How many times larger is the diameter of the largest planet than the diameter of the smallest?

2. Rank the planets (classical planets, dwarf planets, and plutoids) from largest to smallest.

3. Describe in detail how the Earth compares to the Sun, in terms of size.

4. Describe two of the most important observations you made when you brought your collection of planets near the model Sun.

5. Watch the following four-minute clip, featuring Neil DeGrasse Tyson being confronted by Pluto (*http://goo.gl/laa087*). Write one "I learned …" statement.

6. Go to the Scale of the Universe website (*http://htwins.net/*scale2) and spend 10 minutes exploring its features and comparisons. Write one "I learned …" statement.

7. Watch this quick video of what it would look like if our Moon were replaced with different planets: (*http://goo.gl/iqjZy7*). Write one "I learned …" statement.

8. (Enrichment) Measure the circumference of our model Sun. Use the formula, $D = C \div \pi$, to calculate its diameter, where D is the diameter, C is the circumference, and π is the constant 3.14. Is the value for the Sun's diameter, which you calculated here, the same value as that obtained in procedure #1? Explain any differences.

9. (Enrichment) Calculate the relative volumes of all of the planets and the Sun, compared with the volume of Earth. Actual diameters of solar system objects are: Sun 1,392,000 km; Mercury 4,878 km; Venus 12,102 km; Earth 12,756 km; Mars 6,794; Ceres 940 km; Jupiter 142,984 km; Saturn 120,536 km; Uranus 51,118 km; Neptune 49,528 km; Pluto 2,300 km; Charon 1276 km; and Eris 2,400 km. How do your calculations compare with the statement that the Sun contains more than one million times as much volume as the Earth?

Learning Target

Determine what objects are included in our solar system.

I Learned:

Redo:

Manipulated Variable:

Measured Variable:

Controlled Variable:

2A. Sizing up the Solar System Lab

NGSS Alignment

MS-ESS1-2. Develop and use a model to describe the role of gravity in the motions within galaxies and the solar system.

MS-ESS1-3. Analyze and interpret data to determine scale properties of objects in the solar system.

2B.

Keeping Your Distance Lab

Problem

How far apart are the planets?

Prediction

Describe how big you think the solar system is, in terms of distance between the planets.

(*Teacher note:* Have students share several predictions out loud, so misconceptions can be anticipated and explained.)

Thinking About the Problem

Have you ever thought about how big our solar system is? With all of the classical planets, the plutoids, the dwarf planets, their moons, and the many comets and asteroids you are learning about, it may seem pretty crowded. But space in the solar system is surprisingly empty. Planets, which are the largest bodies orbiting the Sun, are tiny compared with the Sun itself.

The distances between the planets are immensely larger than the size of the planets themselves. You may already know that Earth orbits the Sun at an average distance of 150 million kilometers (93 million miles). This distance is called one astronomical unit, or 1 AU. But do you know how far Mars is from the Sun? Do you know how far Neptune is?

Beyond Neptune is a region of icy objects, some with very large diameters, orbiting the Sun in what is called the trans-Neptunian region. The innermost section of this faraway region is the Kuiper belt, named after Dutch-American astronomer Gerard Kuiper. It is in this region where most plutoids, such as Pluto and Eris, and some dwarf planets, such as Charon, are found. It is also believed that this region could be the source for some comets, such as Comet Halley. Our solar system stretches out to 50 AU from the Sun. We have found at least two more dwarf planets, including Sedna, and many comets beyond that distance, however, in the

icy Oort cloud surrounding our solar system (named after Dutch astronomer Jan Oort).

In this two-part lab, you will see for yourself how widely separated the planets are. While we will use dots along a line in order to illustrate the orbits of the planets, their actual orbits are elliptical (*elleipein* "to fall short" in Greek). Elliptical orbits are slightly flattened circles. Solar system objects could be found anywhere along this elliptical orbit. We place them in a line for learning purposes only.

Due to the elliptical shape of the orbit, the distance is actually an average. At *perihelion* (the closest that our planet comes to the Sun), in the beginning of January, the distance is 0.98 AU. At *aphelion* (the farthest that our planet is from the Sun), in the beginning of July, the distance is 1.02 AU.

Write three main points from the "Thinking About the Problem" reading:

1.

2.

3.

Procedure

1. Use the ruler to place a dot 5 cm from the left edge of your one-meter-long Paper Strip Solar System. Label the dot "Sun."
2. Refer to Data Table 2.2 (p. 50) for the orbit distance in centimeters. (Use the scale of 1 AU = 1 cm). Label a dot for each solar system object, following the required scale, as you go outward in a line from the Sun. Remember that typically, the planets would never line up this way, however.
3. Draw and label a sketch of the planets and the Sun in the box on p. 49, showing approximate distances. Your sketch should show the approximate the spacing between planets.

DRAW LABELED SKETCH FOR PROCEDURE #3 HERE.

Learning Target

Develop a better understanding about how widely separated the planets are in our solar system.

I Learned:

Redo:

Manipulated Variable:

Measured Variable:

Controlled Variable:

Analysis

1. Describe the trend, or pattern, found in the distance between planets as you move outward from the Sun.
2. Combine what you've learned from both labs (*Keeping Your Distance* and *Sizing Up the Solar System*). What general statement can you make about how the size of a planet relates to its distance from the Sun?
3. Describe one main quality about each of the following groups of planets:
   a. Terrestrial planets:
   b. Gas giant planets:
   c. Ice giant planets:
4. In order to be called a "dwarf planet," what qualities must a solar system object have?
5. (Enrichment) Given that the circumference of Earth at the equator is 40,000 km and the time for one rotation is 24 hours, what is the rotation speed at the equator in km/hr?
6. (Enrichment) Given that the distance from the Sun is 150,000,000 km (radius of ellipse), and the time for one revolution is 365.25 days (calculate that in hours), calculate the circumference of Earth's orbit ($C=2\pi r$) in km.

DATA TABLE 2.2.

DISTANCES BETWEEN THE SUN AND PLANETS

PLANET NAME	AVERAGE DISTANCE (AU)
Classical Planets	
Mercury	0.39
Venus	0.72
Earth	1.00
Mars	1.50
Jupiter	5.20
Saturn	9.50
Uranus	19.20
Neptune	30.10
Plutoids	
Pluto	39.50
Eris	97.00
Dwarf Planets	
Ceres	2.80
Charon	39.50
Haumea	43.10
Quaoar	43.60
Makemake	45.80
Sedna	975.00

7. (Enrichment) Given the circumference of Earth's orbit in question #3, and that $v = d/t$, calculate the revolution speed around the Sun in km/hr.
8. (Enrichment) Visit the following three websites and then write an "I learned…" statement regarding each. Since *FlashPlayer* is required for some, you may need a different computer than your iPad for this assignment.
    - If the Moon were only one pixel: *http://joshworth.com/dev/pixelspace/pixelspace_solarsystem.html*
    - Astronomical distance conversion: *http://convertxy.com/index.php/astronomicals*
    - Explore the Solar System: *http://bobs-spaces.net/explore-the-solar-system*
9. (Enrichment) Make an Our Community Solar System Map. Obtain a screen shot of a roadmap of our community, with our school in the middle of the map. Label our school, "Sun." Using a scale of 1 cm = 1 AU, draw a circle representing each planet's orbit at the appropriate distance from the Sun. Find and label your home to show which planet orbits nearest it.
10. (Enrichment) Approximately how many times farther from the Sun is Pluto than Earth? How about the dwarf planet, Charon? How about the plutoid, Eris? Describe the distances involved in the other objects found in the Kuiper belt.

Walk Through the Solar System Worksheet

During class today, you will be making a journey through our solar system, starting at our closest star, the Sun. Along the way, you will learn about all of the major solar system objects, including planets, moons, and dwarf planets. Our journey is made according to the scale of the actual distances, but remember that planets can be anywhere along their orbital journey around the Sun. In this activity, the distance you travel represents the average distance from the Sun for each particular object.

Throughout the journey, keep track of your location on the map shown in Figure 2.2 (p. 52). For each stop, plot the point on the map where that solar system object can be found, and label the name of the object, represented by that point. Then write one detailed, new concept that you learned from the brief video at each stop. Turn the sound completely off on your iPad, as the videos are picture and text only, so you can watch at your own pace.

FIGURE 2.2.

SAMPLE MAP OF SOLAR SYSTEM OBJECTS IN SCHOOL BUILDING

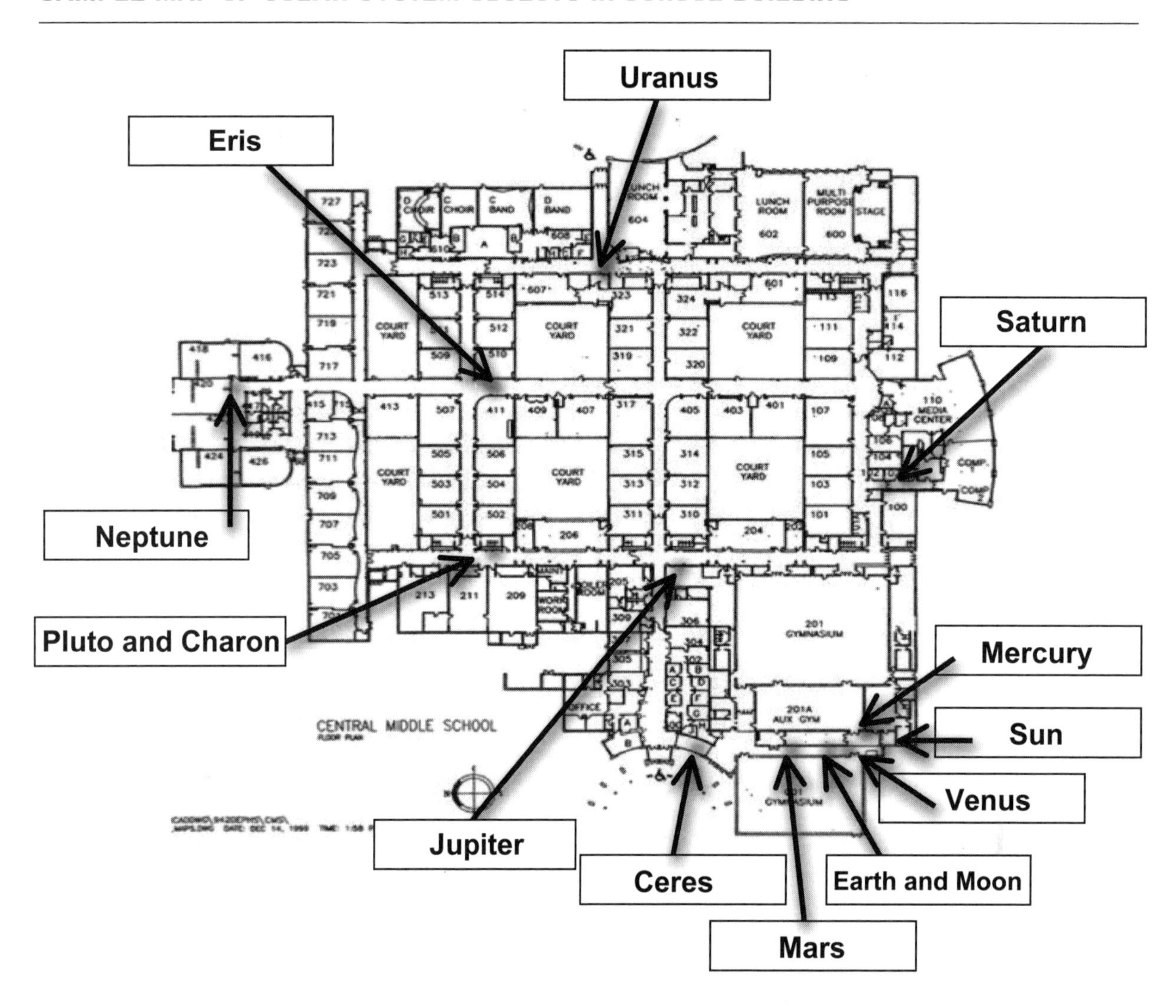

After plotting the point on the map where that Solar System Object can be found, write one detailed, new concept that you learned from the brief video at each stop. Use the *Scan* app on your iPad to view the QR Code for each video

1. Solar System and the Sun

www.youtube.com/watch?v=TtF6H-757cw

2. Mercury

www.youtube.com/watch?v=bACaYC5f7eA

3. Venus

www.youtube.com/watch?v=CqTeqxCTxQ4

4. Earth

www.youtube.com/watch?v=zVbepmvd3QY

5. Earth's Moon

www.youtube.com/watch?v=NgvSVkGLaLA

6. Mars

www.youtube.com/watch?v=tmRXEKkDQNs

7. Ceres and The Asteroid Belt

www.youtube.com/watch?v=9e9Now90Ux8

8. Jupiter

www.youtube.com/watch?v=Ln6-faoni_c

9. Saturn

www.youtube.com/watch?v=FusDAwS8DsY

10. Uranus

www.youtube.com/watch?v=4pPLdUxlReI

11. Neptune

www.youtube.com/watch?v=XTmlKa8CwJE

12. Pluto and Charon

www.youtube.com/watch?v=kdk-pPPRlCo

13. Eris

www.youtube.com/watch?v=ypycJAIXerE

2B. Keeping Your Distance Lab

NGSS Alignment

MS-ESS1-2. Develop and use a model to describe the role of gravity in the motions within galaxies and the solar system.

MS-ESS1-3. Analyze and interpret data to determine scale properties of objects in the solar system.

2C.

Reflecting on the Solar System Lab

Problem

Can the amount of reflected light teach us about a distant planet?

Prediction

Describe, in one sentence, what materials you would use to build a small object that would increase in temperature when placed in direct sunlight.

(*Teacher note:* Have students share several predictions out loud, so misconceptions can be anticipated and explained.)

Thinking About the Problem

Have you ever looked up at the Moon on a cloudless night? The Moon seems so bright; you might imagine it to be covered in something highly reflective, such as snow or bright yellow paint. But looks are deceiving. In fact, if you held a piece of the lunar surface (*luna* "moon" in Latin and Spanish, *lune* in French) in your hand it would appear very dark. The Moon only appears bright in the sky because it is illuminated by a tremendous amount of light from the Sun. It reflects some of that light back out into space, enabling you to see it so clearly.

The fraction of the light falling on a solar system object that is then reflected back into space is called its *albedo*. The albedo (*albus* "white" in Latin) of an object can range from nearly zero (no light reflected back) to almost one (all light reflected back).

Observing the albedo of an object can help us determine what materials make up the object. A low albedo probably indicates a surface composed of dark rocks, as on the Moon. A high albedo is often due to the presence of clouds, as on Venus and Jupiter, or of a frozen icy surface, as on Neptune.

Earth's albedo is less predictable because of large variations in cloud cover. The wide variation in the Earth's albedo is an important factor in studies of long-term climate and global warming. Any light energy that does penetrate the clouds gets

trapped below them, a condition called the greenhouse effect. A low albedo means that much of the incoming light and heat energy from the Sun is absorbed, rather than reflected. The object that absorbs most of this energy will be warmer than an object that reflects away most of its light energy.

Write three main points from the "Thinking About the Problem" reading:

1. Low albedo means ...

2.

3.

Procedure

1. Build a low albedo device (LAD) out of simple materials.
2. Examine your LAD. Make mental observations about its size, color, and construction materials. Insert a labeled picture of your LAD (box on p. 59).
3. Magnifying glasses (and heat sources) will not be allowed, as this assignment is about reflectivity, and your device must hold one metric thermometer.
4. Record the temperature (in °C) versus time (in minutes) results of your LAD investigation for 20 minutes in Data Table 2.3 (p. 61).
5. Use the information in Data Table 2.4 (p. 62) to organize and compare the list of albedos of various solar system objects for Data Table 2.5 (p. 63).

Analysis

1. Give a working definition of *albedo*.
2. Compare your LAD with your classmates' LADs. Which method worked best for reflecting the least light energy? Explain.
3. Determine, using the Gray Scale chart, what the average albedo is for your LAD.

INSERT LABELED IMAGE FOR PROCEDURE #2 HERE.

4. Which solar system object in Data Table 2.4 has the highest albedo? Which has the lowest albedo?
5. Is the albedo of the Earth's Moon more like that of a snowball or a lump of charcoal?
6. Describe, in detail, the changes in temperature of your LAD during the 20 minutes in the Sun.
7. How did the albedo of your device influence what happened during the 20 minutes in the Sun?
8. (Enrichment) The albedo of Earth is such a critical factor in understanding climate and global warming that it has become a very important quantity to measure. Use resources to learn about specifics on Earth's albedo. In one detailed paragraph, describe specifically how the Earth's albedo changes as you move around the surface of the Earth.

9. (Enrichment) Use the following link to investigate how Earth's magnetosphere impacts our climate, environment, and atmosphere (and therefore our albedo): *http://poleshift.ning.com/profiles/blogs/real-time-magnetosphere-data-reading-between-the-lines*. Write one detailed paragraph about your findings.
10. (Enrichment) How would a spectroscope enrich our understanding of albedo?
11. (Enrichment) Make an iMovie trailer that describes specifically how Earth's albedo changes with the seasons. Include all necessary details.
12. (Enrichment) Use the information collected in Data Table 2.3 to make a graph of temperature versus time. Find an equation to describe the line generated by this data.

Learning Target

Determine how albedo helps us learn about solar system objects.

I Learned:

Redo:

Manipulated Variable:

Measured Variable:

Controlled Variable:

DATA TABLE 2.3.

TIME VS. TEMPERATURE FOR LAD

TIME (MIN.)	TEMP. (°C)	TIME (MIN)	TEMP. (°C)
0.0		10.5	
0.5		11.0	
1.0		11.5	
1.5		12.0	
2.0		12.5	
2.5		13.0	
3.0		13.5	
3.5		14.0	
4.0		14.5	
4.5		15.0	
5.0		15.5	
5.5		16.0	
6.0		16.5	
6.5		17.0	
7.0		17.5	
7.5		18.0	
8.0		18.5	
8.5		19.0	
9.0		19.5	
9.5		20.0	
10.0			

DATA TABLE 2.4.

ALBEDOS OF VARIOUS SOLAR SYSTEM OBJECTS

OBJECT	TYPE OF OBJECT	ALBEDO
Mercury	Planet	0.06
Venus	Planet	0.76
Earth	Planet	0.4 (average)
Moon	Moon of Earth	0.07
Mars	Planet	0.16
Phobos	Moon of Mars	0.018
Ceres	Dwarf planet (former asteroid)	0.11
Vesta	Asteroid	0.38
Jupiter	Planet	0.51
Europa	Moon of Jupiter	0.6
Callisto	Moon of Jupiter	0.2
Saturn	Planet	0.50
Titan	Moon of Saturn	0.20
Uranus	Planet	0.66
Oberon	Moon of Uranus	0.05
Neptune	Planet	0.62
Triton	Moon of Neptune	0.80
Pluto	Plutoid (trans-Neptunian object)	0.5
Charon	Dwarf Planet (former moon of Pluto)	0.37
Eris	Plutoid (trans-Neptunian object)	0.75

DATA TABLE 2.5.

PLANETARY ALBEDOS WITH GRAY SCALE CHART

GRAY SCALE CHART	ALBEDO	SOLAR SYSTEM OBJECTS
	0.00–0.10	
	0.11–0.30	
	0.31–0.45	
	0.46–0.55	
	0.56–0.85	
	0.86–1.00	

2C. Reflecting on the Solar System Lab

NGSS Alignment

MS-ESS1-1. Develop and use a model of the Earth-Sun-Moon system to describe the cyclic patterns of lunar phases, eclipses of the Sun and Moon, and seasons.

MS-ESS1-3. Analyze and interpret data to determine scale properties of objects in the solar system.

MS-ETS1-1. Define the criteria and constraints of a design problem with sufficient precision to ensure a successful solution, taking into account relevant scientific principles and potential impacts on people and the natural environment that may limit possible solutions.

MS-ETS1-2. Evaluate competing design solutions using a systematic process to determine how well they meet the criteria and constraints of the problem.

MS-ETS1-3. Analyze data from tests to determine similarities and differences among several design solutions to identify the best characteristics of each that can be combined into a new solution to better meet the criteria for success.

MS-ETS1-4. Develop a model to generate data for iterative testing and modification of a proposed object, tool, or process such that an optimal design can be achieved.

5-ESS1-1. Support an argument that differences in the apparent brightness of the Sun compared to other stars is due to their relative distances from the Earth.

4-PS4-2. Develop a model to describe that light reflecting from objects and entering the eye allows objects to be seen.

2D.

Comparing Planetary Compounds Lab

Problem

How are the planets that are closer to the Sun different from the planets that are farther away?

Prediction

Describe, in one sentence, how you think the planets differ from each other.

(*Teacher note:* Have students share several predictions out loud, so misconceptions can be anticipated and explained.)

Thinking About the Problem

What are the planets made of? Isn't the firm ground we stand on basically the same as it would be on any of the other planets? Well, there are some planets that we would have a very hard time "standing" on, because they are composed basically of compressed gases and ice, and their vast size makes their gravitational pull far too strong for human legs to withstand. These include the biggest of the planets in our solar system: Jupiter, Neptune, Saturn, and Uranus.

Made up mainly of lighter compounds, such as methane and hydrogen, the *gas giant* planets are only really solid at their core. Conversely, the four planets that are closest to the Sun, Earth, Mars, Mercury, and Venus, are solid both at their core and at their crust. Because they each have a surface that is hard, they are grouped together as the terrestrial planets (*terra* in Latin, *terre* in French, and *tierra* in Spanish all mean "Earth"). The basic composition of Earth's core is iron, but there are also significant amounts of nickel and sulfur.

The composition of the planets is dictated in part by their distance from the Sun. Separating the gas giants from the terrestrial planets is a frost line, located 3.4 AU from the Sun, in the main asteroid belt. Inside the frost line, temperatures were high enough that only metallic elements were able to condense into solids. Beyond

the frost line, temperatures were cool enough to allow even hydrogen and helium to condense into solid ices. By comparing the gas giants and the terrestrial planets we can gain insight into the formation of our solar system.

At its start, our solar system was formed with a young star, that we call Sun, plus a considerable amount of matter within the Sun's gravitational reach, including very hot metals and gasses. As time went by, this matter cooled and rocks and metals that were closer to the Sun condensed into solids, while gasses did not. In the outermost planets, however, the light compounds as well as the rocks and metals condensed into types of solid ice. On the terrestrial planets, light compounds such as hydrogen and methane remain as gasses; while on the gas giant planets, being much farther from the Sun's warmth, even these light compounds often condensed into solid form.

As a result of their relative distance from the Sun, Earth, Mars, Mercury, and Venus are composed mainly of metal and rock and are therefore smaller than the larger planets that contain a lot of ice and gas in addition to their solid core. In this lab, you will investigate densities of ice, rock, and iron, in order to learn about the composition of the planets.

Write three main points from the "Thinking About the Problem" reading:

1.

2.

3.

DRAW LABELED SKETCH OF EXPERIMENTAL MATERIALS HERE.

FIGURE 2.3.

MENISCUS DIAGRAM

Here is a simple meniscus diagram, showing this adhesion-of-water-in-a-cylinder event. Reading the volume from the bottom of the meniscus enables you to be more accurate. Without an accurarate volume, your density calculations will be incorrect.

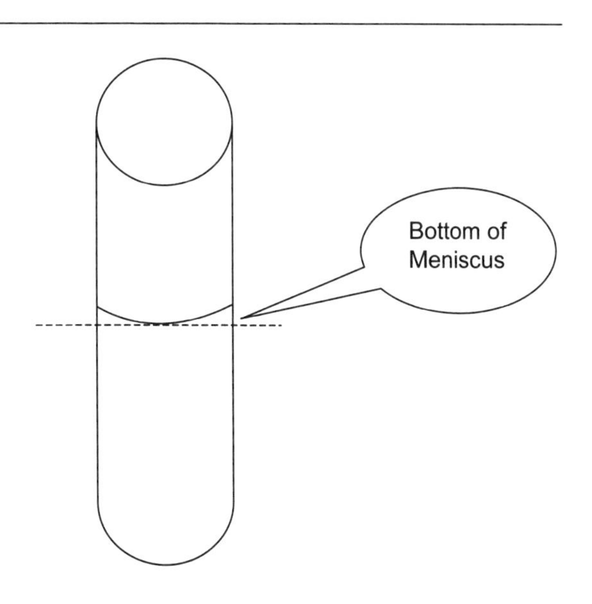

Procedure

1. Fill the graduated cylinder with about 100 ml of water. Read the exact volume (see Figure 2.3, p. 67) and record this value as the "starting water volume" in Data Table 2.6.
2. Measure the mass of the rock in grams. Repeat and get an average. Record results in Data Table 2.6.
3. As soon as you have measured the rock's mass, place it into the graduated cylinder. Read the exact "ending water volume" and record it in Data Table 2.6.
4. Subtract the starting water volume from the ending water volume, entering the result as "change in volume" in Data Table 2.6.
5. Remember that density equals mass divided by volume. Use the following equation to calculate the density of the rock:
   Density = Mass of Object (g) ÷ Change in Volume (ml) = Density (g/ml)
6. Repeat steps 1–5, using the iron and then the ice. When working with the ice, work quickly. Since it floats, use the tip of your pencil to push it down so that it is just barely, yet completely, submerged. Record your results in Data Table 2.6.

DATA TABLE 2.6.

DENSITIES OF PLANET COMPONENTS

OBJECT	ROCK	IRON	ICE
Mass (g)			
Starting water volume (ml)			
Ending water volume (ml)			
Change in volume (ml)			
Density (g/ml)			

7. Use the data from Data Table 2.6, to make a vertical line 2 cm high for rock, iron, and ice on Data Table 2.7.
8. Make a vertical line 1 cm high on Data Table 2.7 for each of the planets listed in Data Table 2.8. Label the lines for each planet. Earth has been done for you.

DATA TABLE 2.7.

DENSITY COMPARISON (G/ML)

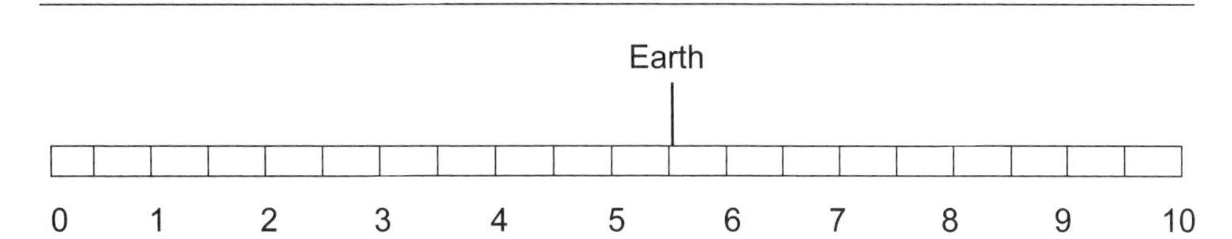

DATA TABLE 2.8.

KNOWN DENSITIES OF PLANETS

CLASSICAL PLANET	DENSITY (G/ML)
Mercury	5.43
Venus	5.24
Earth	5.52
Mars	3.93
Jupiter	1.33
Saturn	0.69
Uranus	1.32
Neptune	1.64

Analysis

1. List rock, iron, and ice in order of increasing density.
2. Given a sample of unknown material, describe one way you might determine if it is more similar to rock, iron, or ice.
3. Looking at Data Table 2.7, determine the two main types of material (rock, iron, or ice) that compose each of the planets listed below. (In reality, "ice" on planets might take the form of compressed gases, such as methane.)

Mercury = ____________________ and ____________________

Venus = ____________________ and ____________________

Earth = ____________________ and ____________________

Mars = ____________________ and ____________________

Jupiter = ____________________ and ____________________

Saturn = ____________________ and ____________________

Uranus = ____________________ and ____________________

Neptune = ____________________ and ____________________

4. (Enrichment) Use resources to find the density of the Earth's Moon. Find the density of Pluto, as well, to see what its composition might be. Compare and contrast the two.
5. (Enrichment) Using the procedure described in this lab, measure the density of five other classroom materials. Report your findings to the teacher.
6. (Enrichment) Use the internet to investigate the composition of each of the other solar system objects. Note that liquid water (required for life, as we know it) has a density similar to, but slightly higher, than ice. Based on your findings, speculate on what other solar system objects might support life.

Learning Target

Investigate densities of ice, rock, and iron, in order to learn about the composition of the planets.

I Learned:

Redo:

Manipulated Variable:

Measured Variable:

Controlled Variable:

2D. Comparing Planetary Compounds Lab

NGSS Alignment

MS-ESS1-4. Construct a scientific explanation based on evidence from rock strata for how the geologic time scale is used to organize Earth's 4.6-billion-year-old history.

MS-ESS2-6. Develop and use a model to describe how unequal heating and rotation of the Earth cause patterns of atmospheric and oceanic circulation that determine regional climates

MS-ESS3-5. Ask questions to clarify evidence of the factors that have caused the rise in global temperatures over the past century.

2E.

Kepler's Laws Lab

Problem

How does a planet's distance from the Sun relate to the time it takes for it to orbit the Sun?

Prediction

Give your best answer, in one sentence, to the problem question above.

(*Teacher note:* Have students share several predictions out loud, so misconceptions can be anticipated and explained.)

Thinking About the Problem

The first person to correctly describe the motion of the planets around the Sun in mathematical terms was Johannes Kepler (1571–1630). Kepler was able to use the carefully collected data from his mentor, Tycho Brahe, to formulate his reasoning. Through painstaking calculations over many years, Kepler discovered three principles, which we call Kepler's laws. These three laws govern how planets orbit the Sun.

Kepler's first law says that the orbits of the planets are shaped like ellipses, or flattened circles. Kepler's second law describes how a planet changes speed while in its orbit, moving faster when it is closer to the Sun (in winter) and more slowly when it is farther from the Sun (in summer). Kepler's third law describes the relationship between the size of a planet's orbit and the time it takes the planet to complete one orbit around the Sun. Since this law involves learning about the other two laws as well, it is the one we will concentrate on during this lab.

The size of an elliptical orbit is described by its axis, *a* (see Data Table 2.9). The Earth's orbital axis, *a*, is defined to be one astronomical unit (1 AU). The values for the other planets are given in AU, which helps compare them with the Earth.

The time it takes for a planet to complete one orbit, is called orbital period, *P*. Earth's orbital period is one year. The values for the other planets are also given in years, which helps compare them with the Earth.

DATA TABLE 2.9.

ORBITAL PERIOD AND AXIS VALUES FOR PLANETS

PLANET TYPE	ORBITAL PERIOD (*P*) (YEARS)	ELLIPTICAL AXIS (*A*) (AU)	P^2 ($P \times P$)	A^3 ($A \times A \times A$)
Classical Planets				
Mercury	0.241	0.387		
Venus	0.615	0.723		
Earth	1.000	1.000		
Mars	1.88	1.524		
Jupiter	11.8	5.20		
Saturn	29.5	9.54		
Uranus	84.0	19.18		
Neptune	165	30.06		
Plutoids				
Pluto	248	39.44		
Eris	557	97.0		
Dwarf Planets				
Ceres	4.6	2.76		
Charon	248	39.44		
Haumea	285	43.10		
Quaoar	287	43.60		
Makemake	310	45.80		
Sedna	12,000	975		

INSERT LABELED SKETCH FOR PROCEDURE #4 HERE.

Kepler's third law states that the square of the orbital period, *P*, is equal to the cube of the axis, *a*. As an equation, it is $P^2 = a^3$. Isaac Newton was later able to use his law of universal gravity to explain that the centripetal force of gravity is how planets maintain this nearly circular orbit. In this lab, you will use this equation to learn about the important mathematical relationships in outer space.

Write three main points from the "Thinking About the Problem" reading:

1. *Ellipitical orbit* means ...

2.

3.

Procedure

1. Calculate the values for the square of the orbital period and enter them in the spaces provided (Data Table 2.9). (P^2 = period × period)
2. Calculate the values for the cube of the axis and enter them in the spaces provided. (a^3 = axis × axis × axis)
3. For the innermost four planets, round your answers to three decimal places. For the rest of the planets, do not include decimals in your answers at all.
4. For your labeled sketch (box on p. 74), draw a diagram similar to Figure 2.4, which shows the Earth and Sun. Label where Earth would be in both summer and in winter.

FIGURE 2.4.

ELLIPSE DIAGRAM WITH LABELS

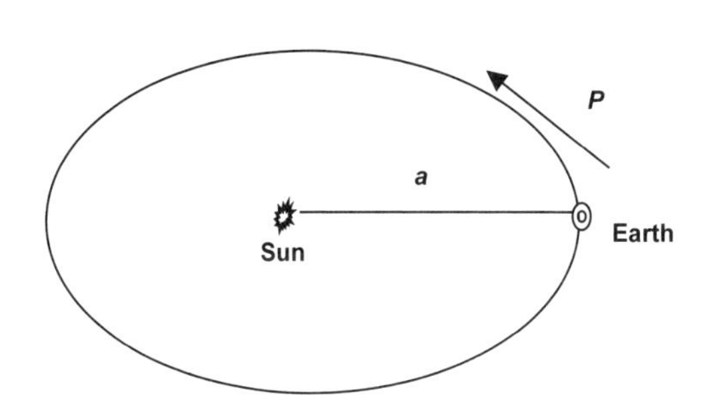

Analysis

1. How do the calculated values in the P^2 and a^3 columns of Data Table 2.9 compare with each other?
2. Is the comparison for Kepler's third law equally good for all of the planets? Describe the similarities and differences.
3. Describe, in detail, how a planet's distance from the Sun relates to the time it takes for it to revolve around the Sun.
4. (Enrichment) When new dwarf planets and plutoids are discovered beyond Pluto, how would their orbital period compare with Pluto? Determine a possible orbital period for an imaginary dwarf planet or plutoid, and then use that to calculate its elliptical axis in AU.
5. (Enrichment) Go to the following website and write one detailed "I learned …" statement: *http://windows.ivv.nasa.gov/the_universe/uts/kepler3.html*

Learning Target

Use the equation for Kepler's third law to learn about mathematical relationships in outer space.

I Learned:

Redo:

Manipulated Variable:

Measured Variable:

Controlled Variable:

2E. Kepler's Laws Lab

NGSS Alignment

MS-ESS1-2. Develop and use a model to describe the role of gravity in the motions within galaxies and the solar system.

2F.

Phasing in the Moon Lab

Problem

Why do we see different Moon phases?

Prediction

Describe, in one sentence, what you think causes the different phases of the Moon.

(*Teacher note:* Have students share several predictions out loud, so misconceptions can be anticipated and explained.)

Thinking About the Problem

Every 29.5 days the Moon goes through a predictable cycle of changes in its shape, which we call phases. For thousands of years, people have recorded these phases, and during this time, the cycle has never changed. Even though it is known what the Moon will look like on any night of the year, many people have misconceptions about why these apparent changes in shape occur. Perhaps the most popular, yet wrong, explanation for Moon phases is that they are caused by Earth's shadow (or eclipse) of the Moon. The truth is actually simpler. Since we are not on the Moon, we can look at it from different angles as it orbits around us. Only the half of the Moon facing the Sun is lit up, and we only get that fully lit view once per month (Full Moon).

Despite seeing different portions of the Moon illuminated in our direction, the physical shape of the Moon never changes. What changes is the Moon's location as it orbits around the Earth. Like the planets, the Moon does not generate its own light. It reflects some of the Sun's light that shines on it. Depending on the Moon's position in its orbit, we see different portions of the Moon's lit side.

The appearance of the Moon goes from a whole circle (Full Moon) to something less than a whole circle (Gibbous Moon), to a half-circle (Quarter Moon), to a Crescent Moon, to almost invisible (New Moon), and back again to the Full Moon. This cycle is called the Lunar Month. When the Moon is approaching full, we call it *waxing*. When it is past full, toward a new Moon, we call it *waning*. In this two-part lab, you will investigate the cause of the phase changes in the Moon.

Write three main points from the "Thinking About the Problem" reading:

1. We see different amounts of the Moon's lit-up side because ...

2.

3.

Procedure (Part A: Whole-Class Investigation)

1. Place the bright lamp in the center of the room. Divide the class in half (January to June birthdays and July to December birthdays).
2. Each student gets one softball (the Moon) and half the class forms a large circle around the lamp.
3. Face the lamp (the Sun) and hold the Moon directly in front of you, slightly above your head.
4. Make observations about the Moon. Notice what portion of the Moon facing you is illuminated by the Sun. It may help to take a photo of what you see with your iPad.
5. Gradually, as directed by your teacher, turn 45° counterclockwise at a time, carefully making observations about what portion of the Moon facing you gets lit up by the Sun.
6. The other half of class now does steps 2–5.

Procedure (Part B: Lab Partner Investigation)

1. Your softball is half-colored in black marker. This simulates the Moon, with the lit-up side facing the Sun and the dark side being harder to see.
2. Stand in an open space of the classroom. Your lab partner should stand 3 meters away, facing you, and holding the Moon at eye level so you can see the dark side only. Pretend the Sun is at the far wall, behind your partner.

FIGURE 2.5.

DATA ON VISIBLE WHITE PORTION OF MOON

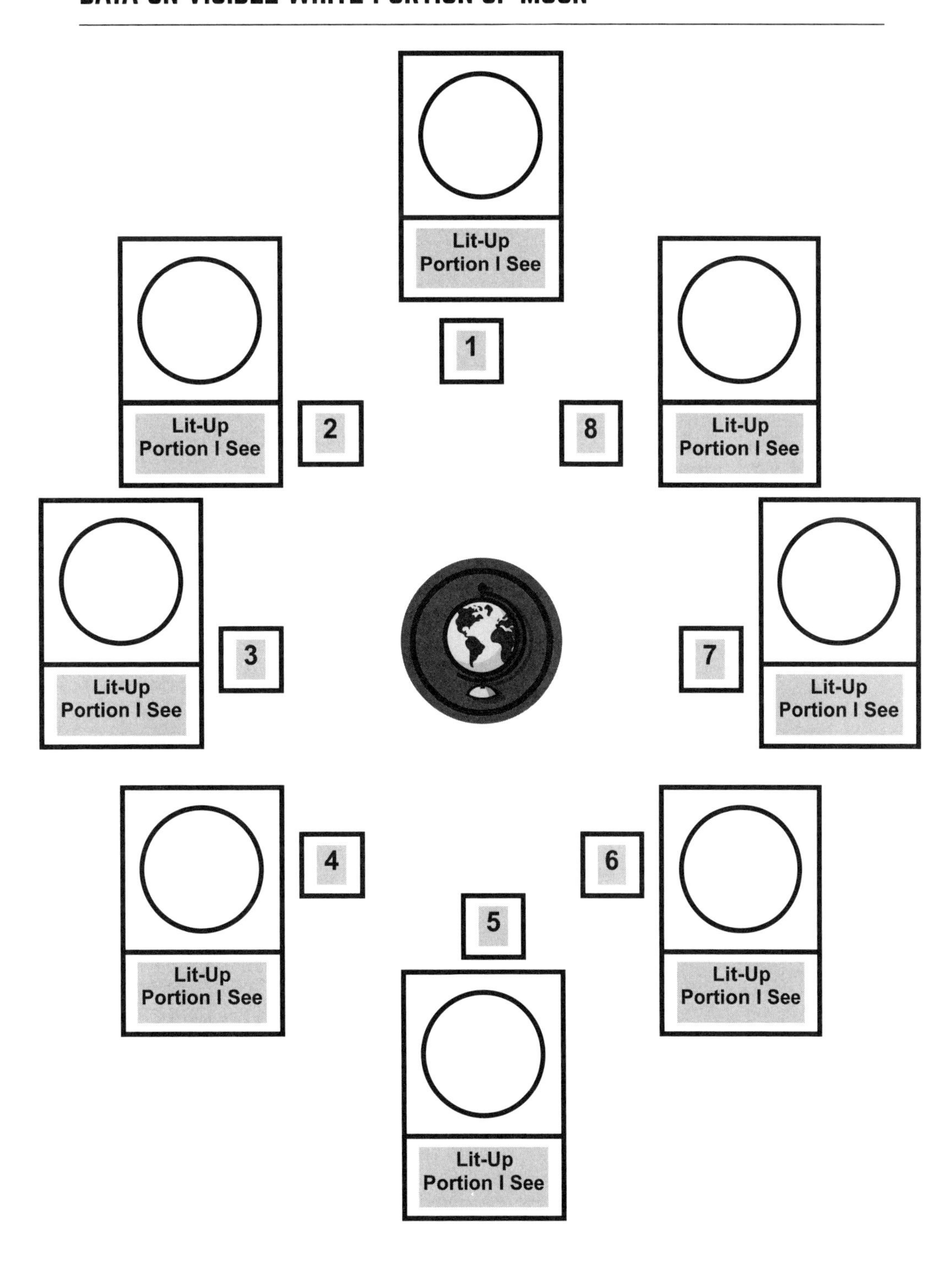

FIGURE 2.6.

CYCLE OF CHANGES ON PORTION OF MOON VISIBLE FROM EARTH

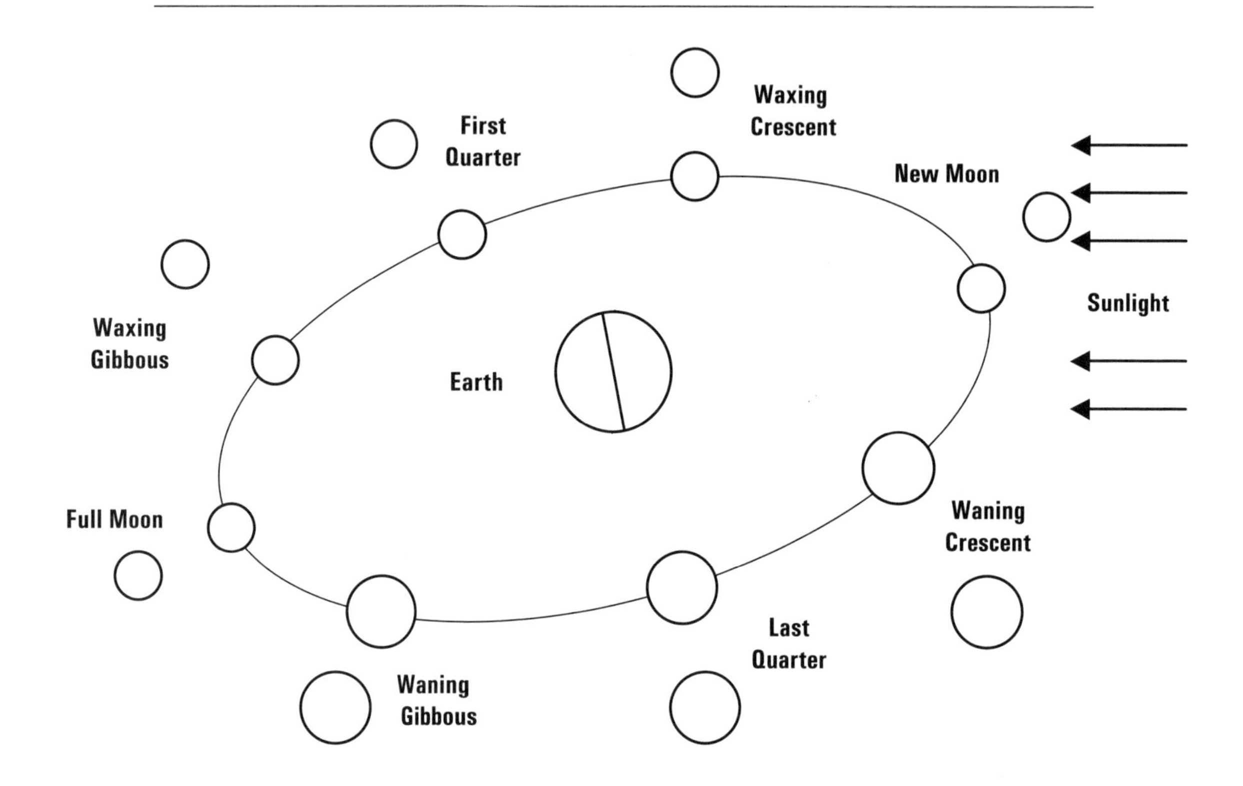

3. The student holding the Moon should move around you in a circle, making sure not to turn the Moon and to always face the same direction (they will have to walk backwards and sideways at times to do this).

4. At each of the eight positions indicated on Figure 2.5 (p. 79), draw only the portion of the white part of the Moon that you can see.

5. Switch places with your lab partner and repeat steps 1–4 to complete your data table.

6. Complete Figure 2.6 with your teacher as a class.

Learning Target

Investigate the cause of the phase changes in the Moon.

I Learned:

Redo:

Manipulated Variable:

Measured Variable:

Controlled Variable:

Analysis

1. Whether or not you could see it in Part A, how much of the Moon's surface area was always illuminated?
2. What fraction of the Moon in Part B, was white (illuminated by the Sun) during the entire activity?
3. Describe, in detail, what happened to the lit-up portion of the Moon that you could see as the Moon went around you in Part B.
4. Describe, in detail, what causes the phases of the Moon to occur.
5. (Enrichment) Although the Sun is 400 times bigger than the Moon, it is also 400 times farther away. This is why both appear (from Earth) to be almost the same size. This is also why the relatively small Moon can completely block our view of the Sun. Use resources to learn about eclipses. Since solar and lunar eclipses occur rarely, what must be true about the Moon's orbit relative to the Earth's orbit during an eclipse?
6. (Enrichment) Explore the following website: *http://aa.usno.navy.mil/data/docs/RS_OneDay.php* and write one "I learned …" statement.

2F. Phasing in the Moon Lab

NGSS Alignment

MS-ESS1-1. Develop and use a model of the Earth-Sun-Moon system to describe the cyclic patterns of lunar phases, eclipses of the Sun and Moon, and seasons.

5-ESS1-2. Represent data in graphical displays to reveal patterns of daily changes in length and direction of shadows, day and night, and the seasonal appearance of some stars in the night sky.

2G.

Reason for the Seasons Reading Guide and Background Reading

Reason for the Seasons Reading Guide Directions

1. Before reading the article, please read the following four statements and indicate whether you agree or disagree with each. In the "Why?" box, then write why you agree or disagree with the particular statement.
2. After reading the article and completing the associated work, read the four statements again, indicating whether you agree or disagree with each now. In the "Why?" box, and then write a new reason why you agree or disagree.
3. Play the Panel of Five Game as you read the background reading.

Statement #1: Earth is closest to the Sun in the summer.	
(Circle)	
Agree	**Disagree**

Statement #2: The orbit of the Earth looks more like a car racetrack than a perfect circle.	
(Circle)	
Agree	**Disagree**

Statement #3: The North Pole always points toward Polaris (the North Star).	
(Circle)	
Agree	**Disagree**

Statement #4: The seasons are caused by the tilt of the Earth's axis.	
(Circle)	
Agree	**Disagree**

Reason for the Seasons Background Reading

The reason why Earth has four seasons—spring, summer, fall, winter—is often misunderstood. In part, this misunderstanding comes from diagrams showing Earth's orbit around the Sun. Having learned that Earth's orbit is an ellipse (although the ellipse is so round it is almost a circle), many people jump to the incorrect conclusion that the planet is hotter in the summer because Earth is closer to the Sun than in the winter.

If the seasons were just caused by Earth's orbit around the Sun, people who live north of the equator and people who live south of the equator would have the same seasons. They do not. For example, if winter occurred because Earth was far away from the Sun (which it does not), then every place on Earth would be cold at the same time. It is not.

In fact, Earth is actually farther away from the Sun during our (the Northern Hemisphere's) summer than it is during our winter. Consider the following example. At the exact same time it is summer in the United States (the Northern Hemisphere), it is winter in Australia (the Southern Hemisphere). If Earth's distance from the Sun determines the seasons, then it should be the same season everywhere on the planet, a condition that never exists.

Earth's orbit around the Sun is more accurately called "slightly elliptical." An ellipse is a flattened circle. It resembles a perfect circle more closely than it does an oval, however. It is only slightly wider than it is tall. This ellipse does not account for seasons. A different characteristic of Earth's motion does, however, explain the seasons.

Earth is spinning. When a ball spins, the line around which the ball turns is called the *axis of rotation*. Earth's axis of rotation is tilted in relation to the plane of Earth's orbit around the Sun. It is tilted at 23.5°. As Earth is spinning, it is also moving (revolving) around the Sun. The position of Earth in its orbit around the Sun, combined with the tilt of the Earth's axis of rotation, determines the season. As Earth revolves around the Sun, its axis always remains tipped in the same direction, toward the distant North Star, Polaris. This means that when the top half of the earth is pointed toward the North Star, the Northern Hemisphere is tipped away from the Sun, and the Southern Hemisphere is tipped toward it. This is why the two hemispheres experience opposite seasons during our yearlong revolution around the Sun.

Austin, Texas is located at latitude 30° and Eden Prairie, Minnesota, is located at latitude 45°. At noon on the summer solstice in June (the day with the longest daylight hours of the year) the Sun's rays strike Austin at 83.5° and strike Eden Prairie at 68.5°. At this moment, the Sun's rays are most intense, and result in a warm summer season. At noon on the winter solstice in December (the day with the shortest daylight hours of the year) the Sun's rays strike Austin at an angle of 36.5° and strike Eden Prairie at 21.5°. In this case, the Sun's rays are hitting the surface at a greater slant than they do during the summer, resulting in less intensity and a cooler winter season.

Lower intensity of sunlight hitting a surface is not as efficient in heating the surface, because the same amount of solar radiation is spread out over a much larger area. When part of the surface of the Earth is tilted with respect to the rays of sunlight, it will receive slanting solar rays and not heat as quickly. Since Earth is round, some locations can be receiving direct sunlight while others receive it indirectly. This explains why it can be summer on one part of the planet and winter on another.

Earth is a very special planet in many ways. Just as Earth's unique atmosphere and its distance from the Sun work together to make Earth the right temperature to support life, Earth's revolution and the tilt of its axis work together to create the seasons.

Using Vocabulary From the Reading

1. Use the Think-Ink-Pair-Share strategy to develop brief definitions for the following list of 10 words from the reading

   - Misunderstanding:
   - Orbit:
   - Ellipse:

- Distance:
- Revolution:
- Equator:
- Hemisphere:
- Rotation:
- Axis:
- Seasons:

2. In small groups, collaboratively compose a detailed paragraph using the 10 words. The words are listed in the same order in which they should appear in your paragraph. There are three rules about the words for this activity:
   a. The words must be used in order.
   b. Once the word has been used, it is okay to use it again.
   c. The word can be used in its various forms (plural, different tenses, and so on).
3. Edit and revise your detailed paragraph or short story, so it is clear, can be read easily by others in the class, and is scientifically accurate.

(*Teacher note:* The author recommends *Kinesthetic Astronomy* by Dr. Cherilynn A. Morrow and Michael Zawaski [*www.space-science.org/education/extra/kinesthetic_astronomy*]. They have an especially good lesson that has students moving through the meaning of *day*, *year*, and *seasons*.)

2G. Reason for the Seasons Reading Guide and Background Reading

NGSS Alignment

MS-ESS1-1. Develop and use a model of the Earth-Sun-Moon system to describe the cyclic patterns of lunar phases, eclipses of the Sun and Moon, and seasons.

MS-ESS3-5. Ask questions to clarify evidence of the factors that have caused the rise in global temperatures over the past century.

5-ESS1-2. Represent data in graphical displays to reveal patterns of daily changes in length and direction of shadows, day and night, and the seasonal appearance of some stars in the night sky.

2H.

Changing Lunar Tides Lab

Problem

How do the tides relate to Moon phases?

Prediction

Describe how often the low-and-high-tide pattern repeats itself. (*Teacher note:* Have students share several predictions out loud, so that misconceptions can be anticipated and explained.)

Thinking About the Problem

Have you ever spent time watching the tide come in or go out? Sitting on a beach and focusing on the tide helps us to realize that there is a definite rhythm to large bodies of water. The pattern repeats itself twice each day. Why?

Tides are the daily changes in water surface height caused by the powerful attraction between the Earth and the Moon. Earth and its Moon are attracted to one another due to gravity (*gravis* "heavy" in Latin). Since Earth has a much greater mass, the effect of Earth's gravity is stronger. But our Moon exerts a strong gravitational force on Earth, in return. This gravitational attraction leads to rhythmic rising and falling of the waterline along the beach. Since water is more easily moved than land, water gets pulled over the land that happens to be facing the Moon. The side of the Earth closest to the Moon will experience higher tides.

Tides are predictable changes in sea level that occur at regular intervals. There is a high tide when sea level has risen to its highest point and a low tide at its lowest. One of the impacts of tides is on ocean shipping. In many locations, ships can only come to shore at high tide. If they come in at low tide, they risk running aground.

Because tides affect us in so many ways, it is important to know when they will occur, and fortunately this can be predicted with accuracy. By making a graph of the tides and the sea level, you can see the pattern of changes over time. The data used in this activity is from the shore of a small island off the coast of Milford, Connecticut, where the author spent many summers. It is located in Long Island Sound on the Atlantic Ocean, across from Port Jefferson, New York.

Write three main points from the "Thinking About the Problem" reading:

1. Tides are ...

2.

3.

Procedure

1. Locate the tide chart for Charles Island in Milford, Connecticut (Data Table 2.10, pp. 88–89). Note that all times are in military (24-hour) time. For example, 0530 refers to 5:30 a.m., and 1522 refers to 3:22 p.m. An average of two low tides (A and C) and two high tides (B and D) occur each day.
2. Study an online reference, so you can draw a labeled sketch (Figures 2.7 A&B, p. 90), which shows the position of the Sun, Earth, and Moon during a high tide.
3. Number the horizontal axis of your graph paper so that you can fit 30 days-worth of data across the bottom (Figures 2.8 A&B, pp. 92–93). The vertical axis will be height of sea level. Give the graph a descriptive title. Write a key describing your color codes.
4. Plot your data for sea level versus time (in days), by first plotting all of the data for "A" (earliest low tide). Draw a color-coded line to connect all of the "A" points in the order that they were plotted.
5. Repeat step #3 by plotting the data and drawing the lines for "B," "C," and "D," Use a different-colored pencil for each line.
6. In the middle of your graph, at approximately the 2.5-meter sea level height, draw the Moon icon, to show what phase it is in on the five particular days.

DATA TABLE 2.10.

TIDE CHART FOR MILFORD, CONNECTICUT

DAY	TIMES	SEA LEVEL (M)	DAY	TIMES	SEA LEVEL (M)
1	A. 0523 B. 1130 C. 1752 D. 2351	A. 0.9 B. 3.7 C. 0.9 D. 4.0	11	C. 0137 D. 0751 A. 1353 B. 1924	C. 0.8 D. 3.6 A. 0.9 B. 3.8
2	A. 0601 B. 1247 C. 1807 D. 2345	A. 0.8 B. 3.7 C. 0.9 D. 3.9	12	C. 0214 D. 0801 A. 1413 B. 2136	C. 0.8 D. 3.6 A. 0.9 B. 3.8
3	A. 0647 B. 1203 C. 1807 D. 2359	A. 0.8 B. 3.6 C. 0.9 D. 3.8	13	C. 0322 D. 0950 A. 1547 B. 2103	C. 0.7 D 3.7 A. 0.8 B. 4.0
4	A. 0728 B. 1316 C. 1905	A. 0.8 B. 3.7 C. 0.95	14	C. 0442 D. 1000 A. 1642 B. 2232	C. 0.8 D. 3.7 A. 0.9 B. 3.9
5	D. 0003 A. 0831 B. 1407 C. 2005	D. 3.7 A. 0.7 B. 3.8 C. 0.5	15	C. 0551 D. 1005 A. 1732 B. 2341	C. 0.9 D. 3.6 A. 1.0 B. 3.8
6	D. 0232 A. 0920 B. 1530 C. 2115	D. 3.8 A. 0.7 B. 3.9 C. 0.6	16	C. 0657 D. 1115 A. 1832 B. 0031	C. 0.8 D. 3.6 A. 0.9 B. 3.8
7	D. 0322 A. 0945 B. 1558 C. 2245	D. 3.7 A. 0.7 B. 3.8 C. 0.5	17	C. 0647 D. 1254 A. 1942	C. 0.7 D. 3.7 A. 0.8
8	D. 0412 A. 1049 B. 1645 C. 2333	D.3.5 A. 0.4 B. 3.9 C. 0.5	18	B. 0140 C. 0705 D. 1638 A. 2052	B. 3.7 C. 0.7 D. 3.8 A. 0.5
9	D. 0536 A. 1154 B. 1705	D. 3.8 A. 0.7 B. 3.9	19	B. 0240 C. 0834 D. 1454 A. 2123	B. 3.8 C. 0.7 D. 3.9 A. 0.6
10	C. 0025 D. 0636 A. 1204 B. 1844	C. 0.8 D. 3.7 A. 0.9 B. 3.9	20	B. 0340 C. 0953 D. 1534 A. 2253	B. 3.5 C. 0.8 D. 3.9 A. 0.6

DATA TABLE 2.10. (*continued*)

DAY	TIMES	SEA LEVEL (M)	DAY	TIMES	SEA LEVEL (M)
21	B. 0220 C. 1037 D. 1614 A. 2313	B. 3.8 C. 0.7 D. 3.9 A. 0.6	26	A. 0249 B. 0850 C. 1432 D. 2042	A. 0.8 B. 3.7 C. 0.9 D. 3.9
22	B. 0458 C. 1138 D. 1723 A. 0052	B. 4.1 C. 0.7 D. 4.0 A. 0.7	27	A. 0351 B. 0854 C. 1530 D. 2110	A. 0.7 B. 3.7 C. 0.8 D. 4.0
23	B. 0505 C. 1239 D. 1802	B. 3.5 C. 0.8 D. 3.9	28	A. 0345 B. 0952 C. 1534 D. 2153	A. 0.8 B. 3.6 C. 0.9 D. 3.8
24	A. 0140 B. 0633 C. 1235 D. 1940	A. 0.8 B. 3.6 C. 0.9 D. 3.8	29	A. 0429 B. 1020 C. 1642 D. 2221	A. 0.8 B. 3.4 C. 0.9 D. 3.5
25	A. 0159 B. 0732 C. 1343 D. 1956	A. 0.7 B. 3.7 C. 0.8 D. 4.0	30	A. 0530 B. 1004 C. 1710 D. 2254	A. 0.7 B. 3.7 C. 0.8 D. 4.0

Analysis

(*Teacher note:* Answers are in parentheses.)

1. About how much time passes in between one low tide "A" and the next low tide "C"? (12 hours)
2. About how much time passes in between one high tide "B" and the next high tide "D"? (12 hours)
3. When did the lowest low tide occur? (Day 8) To which Moon phase does this answer correspond? (New Moon)
4. When did the highest high tide occur? (Day 22) To which Moon phase does this answer correspond? (Full Moon)
5. When did the highest low tide occur? (Day 15) To which Moon phase does this answer correspond? (First Quarter)
6. When did the lowest high tide occur? (Day 29) To which Moon phase does this answer correspond? (Last Quarter)

FIGURE 2.7A.

LABELED SKETCH FOR PROCEDURE #2

FIGURE 2.7B

ANSWER

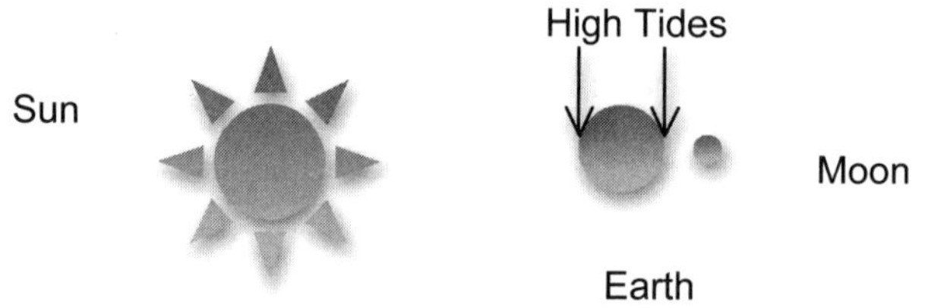

7. Explain the relationship between the Moon's position and the high tide on Earth. (As the Earth spins, sea levels are kept at roughly equal levels around the planet by the Earth's gravity pulling inward and centrifugal force pushing outward. However, the Moon's gravitational forces are strong enough to disrupt this balance by slightly pulling the water toward the Moon. This causes the water to bulge away from Earth, toward the Moon. As the Moon orbits our planet and as the Earth rotates, the bulge moves. The areas of the Earth where the bulging occurs experience high tide, while the other areas are subject to a low tide.)

8. (Enrichment) Long Island Sound, between New York and Connecticut, has shallow sand bars all around it, which prevent even small sailboats from coming close to shore at low tide without being grounded. But at high tide, these sand bars do not cause a problem for boats. On Day 10, the crew of a sailboat wants to get as close as possible to shore, without getting stuck in the sand bars. They radio you at the Coast Guard station to ask what time they should come in, to take advantage of high tide and daylight. What should you tell them? (Approximately 6:36 a.m.)
9. (Enrichment) You are staying at a beach house. At high tide, the water completely covers the part of the beach that is usable. On Day 6, you go out on the beach at 10:00 a.m. to read a book. After an hour the sound of the waves lulls you to sleep. How long can you sleep before you must either wake up or get wet? Explain how you figured this out. (About 4–5 hours. You would start to get wet by about 3:00 p.m.)
10. Explore the following website (*www.noaa.org*) and give one detailed, "I learned …" statement.
11. Explore the following video on the Bay of Fundy and give one detailed "I learned …" statement: *www.youtube.com/watch?feature=player_detailpage&v=l85Dk9LpIEs*

FIGURE 2.8A. ______________________________

(*Teacher note*: Students provide a descriptive title and key for graph.)

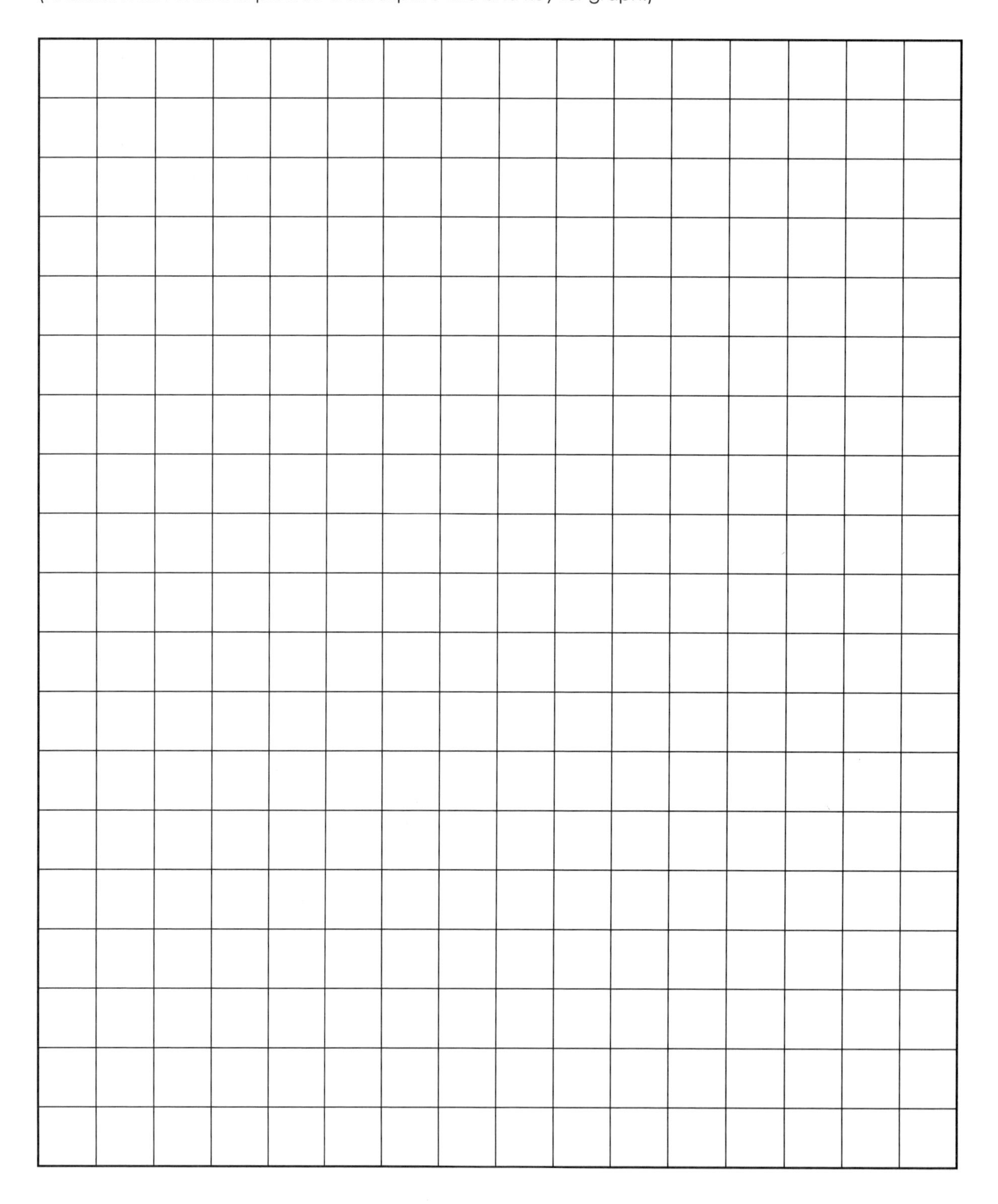

FIGURE 2.8B.

SAMPLE GRAPH OF DATA

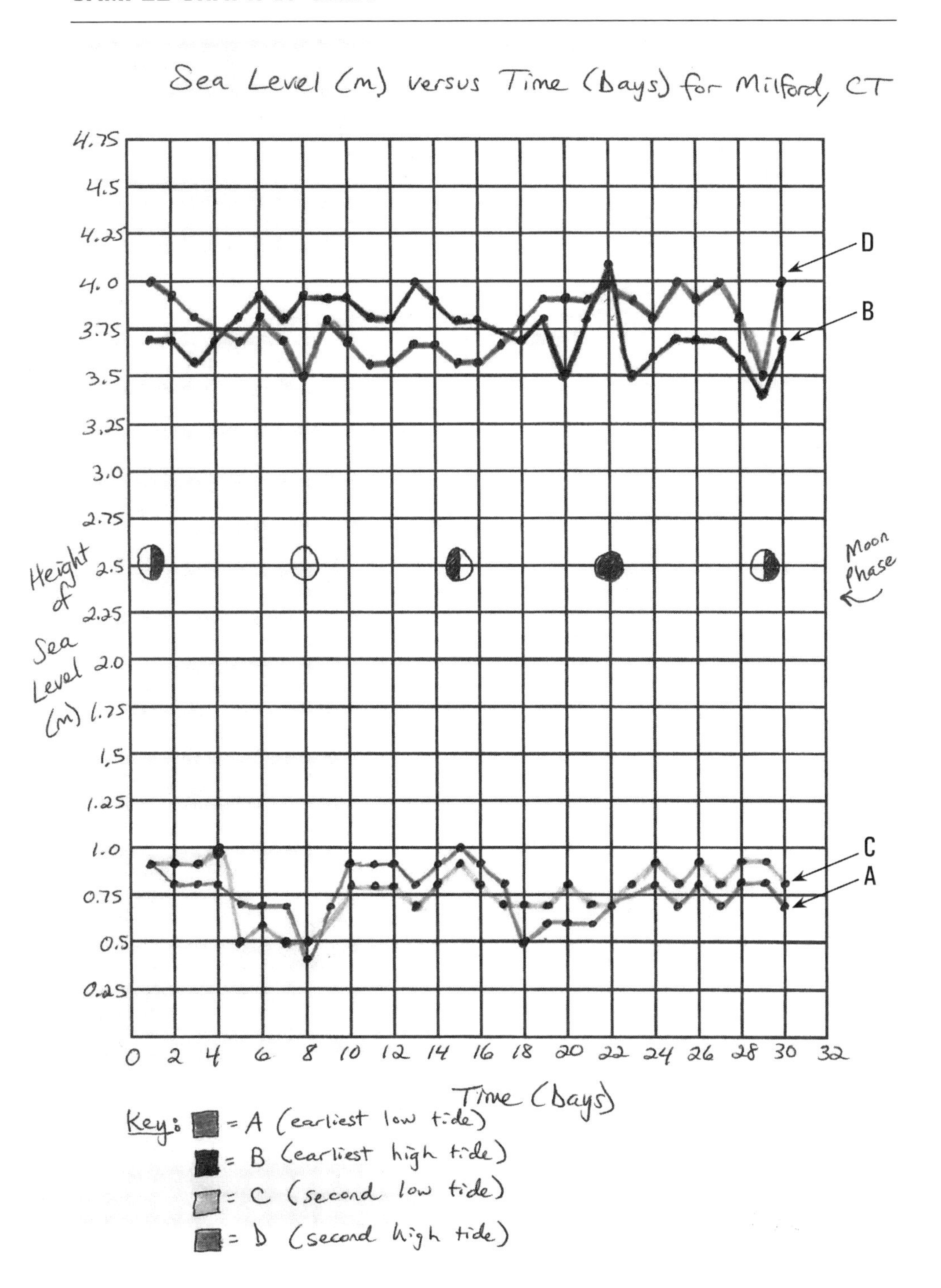

Learning Target

Make a graph of tides and the associated sea levels, to see the pattern of changes over time.

I Learned:

Redo:

Manipulated Variable:

Measured Variable:

Controlled Variable:

2H. Changing Lunar Tides Lab

NGSS Alignment

MS-ESS1-1. Develop and use a model of the Earth-Sun-Moon system to describe the cyclic patterns of lunar phases, eclipses of the Sun and Moon, and seasons.

5-ESS1-2. Represent data in graphical displays to reveal patterns of daily changes in length and direction of shadows, day and night, and the seasonal appearance of some stars in the night sky.

21.

Finding That Star Lab

Problem

Which way will the Big Dipper's handle be facing tonight?

Prediction

Describe, in one sentence, exactly how you think the Big Dipper will look tonight.

(*Teacher note:* Have students share several predictions out loud, so that misconceptions can be anticipated and explained.)

Thinking About the Problem

On a clear night (away from light pollution) you can see lots of stars in the sky. Ancient Greeks, Egyptians, and Native Americans imagined outlines of animals and people as they looked at the stars. Native Americans from the Lakota Tribe have discovered that many of the Petroglyphs in western Minnesota are based on what the tribe's observers knew about the night sky and turned into story format in order to teach it to others. These star outlines were eventually called "constellations."

European astronomers named the constellations based on mythology (legendary stories based on the beliefs of ancient people). The sky was mapped into areas designated by the constellation names. Most of the bright stars also have names based on mythology.

Why do we see constellations? Planet Earth rotates on its axis once each day. Earth's rotation causes not only the Sun, but also all of the other stars, to appear to move across the sky. Earth's rotation is the cause of night and day. Planet Earth also revolves around the Sun once each year. Earth's revolution causes the constellations to appear to move from night to night and month to month.

Star-finding charts help students locate constellations visible at any time for any date. Star finders are also known as *planispheres* because they represent a spherical sky chart on a paper plane. This planisphere lesson was developed with help from

the author's high school Earth Science teacher: D. Louis Finsand of the University of Northern Iowa in Cedar Falls, Iowa.

Write three main points from the "Thinking About the Problem" reading:

1. A planishere is ...

2.

3.

Materials

- Model Earth
- Planisphere Pattern
- Scissors
- Glue
- North, South, East, West Posters

Procedure

1. Each student gets one planisphere pattern (Figures 2.9–2.11, pp. 100–102). Cut out and glue the planisphere as indicated on the pattern. Draw a sketch in the box on p. 98.
2. Follow along with the activity directions given by your teacher.

Teacher Directions to Read Aloud

1. Point to True North. Change the direction of your desk, so that you are facing north.
2. Hold your planisphere at arm's length in front of you. Turn it so that November is at the bottom.
3. Point toward the real horizon outside our classroom. The dashed curve under November represents the horizon on your planisphere.
4. Point to the spot that is directly above your head. The center of the sky overhead is called the *zenith*. The dashed curve at the top of your planisphere represents the zenith.
5. Connect the dots for the Big Dipper. Where is the Big Dipper in the sky during November?
6. Where does the handle point during November?
7. Where does the cup of the Big Dipper point during November?
8. Connect the dots for the Little Dipper, Cepheus, Cassiopeia, and Draco the Dragon.
9. Observe the spinning Model Earth. The spinning of our Earth makes the stars appear to slowly move counterclockwise in our sky. Slowly rotate your planisphere counterclockwise to show what the sky will look like in December. Which constellation is at the zenith in December?
10. Slowly rotate your planisphere to late January. Which constellation is near the horizon?
11. Slowly rotate your planisphere counterclockwise until you see Cepheus is on the northern horizon. During which month does that occur?
12. In April, which way does the cup of the Big Dipper point?

Analysis

1. Where is the North Star located on any night of the year?
2. What constellation will be at our zenith tonight after the Sun sets?
3. What are some of the constellations visible on your birthday?
4. (Enrichment) Explore the following two websites, and write one "I learned …" statement. *www.heavens-above.com* and *www.skyandtelescope.com.*

DRAW LABELED SKETCH OF ASSEMBLED PLANISPHERE HERE.

Learning Target

Use star-finding charts to help students locate visible constellations.

I Learned:

Redo:

Manipulated Variable:

Measured Variable:

Controlled Variable:

DATA TABLE 2.11.

OBSERVATIONS OF PLANISPHERE

PLANISPHERE QUESTION	ANSWER
How do I find the horizon on the plani-sphere?	
What is the zenith?	
How do I find the zenith on the planisphere?	
Where does the handle of the Big Dipper point during November?	
Where does the cup of the Big Dipper point during November?	
Which constellation is at the zenith in December?	
Which constellation is near the horizon in Late January?	
During which month is Cepheus on the northern horizon?	
Which way does the cup of the Big Dipper point in April?	

FIGURE 2.9.

TOP OF SAMPLE PLANISPHERE MODEL

(*Teacher note:* The *Star Walk* app, from *www.apple.com*, can be used. See Figure 2.12 for app details. Credit for these examples goes to *www.csun.edu/science/geoscience/astronomy/planisphere/plansiphere.html*. This link provides directions for assembly, as well.)

FIGURE 2.10.

UNMARKED BASE OF PLANISPHERE MODEL

(*Star Walk* app can also be used.)

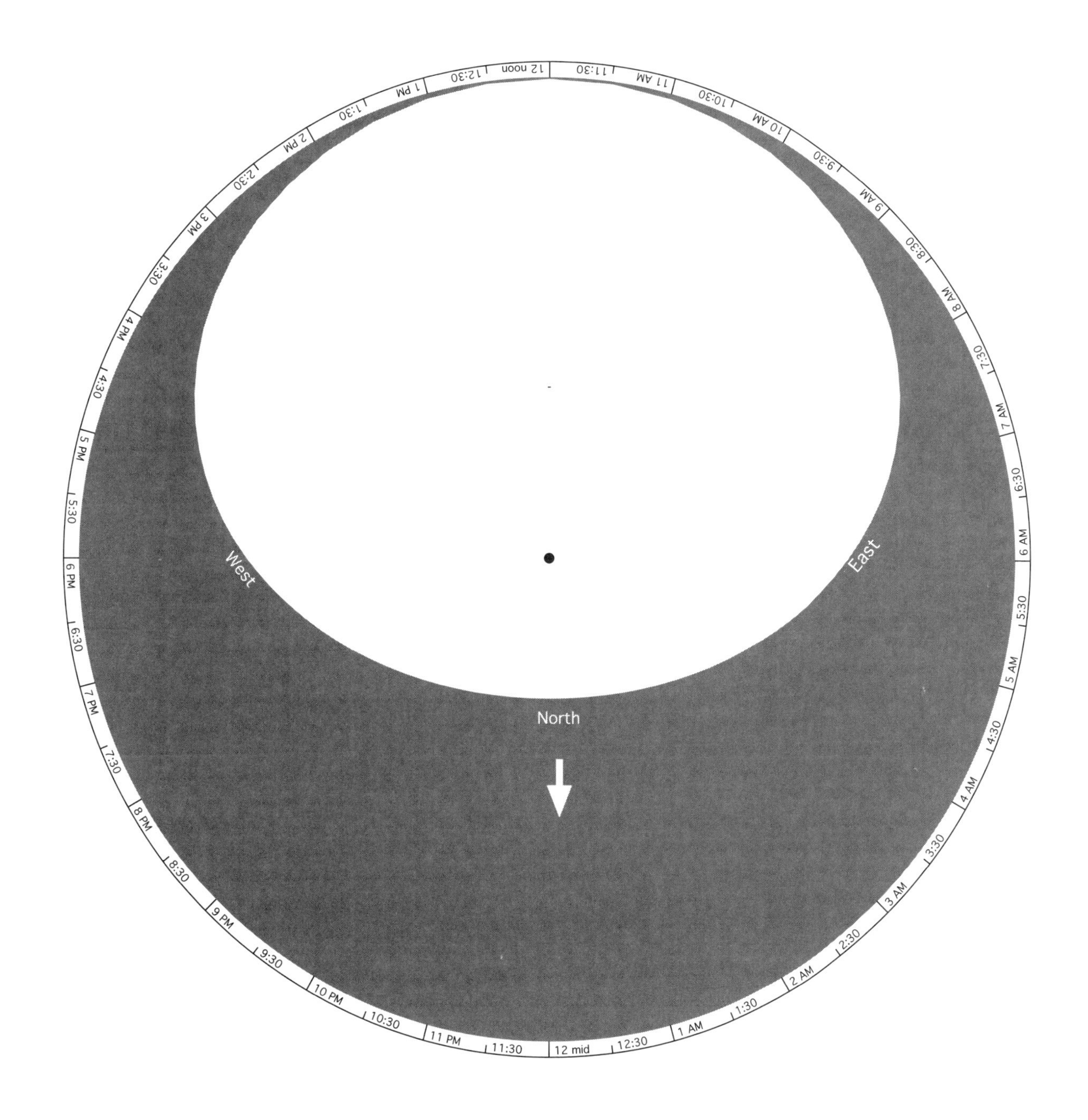

FIGURE 2.11.

BASE OF PLANISPHERE MODEL

(*Star Walk* app can be used.)

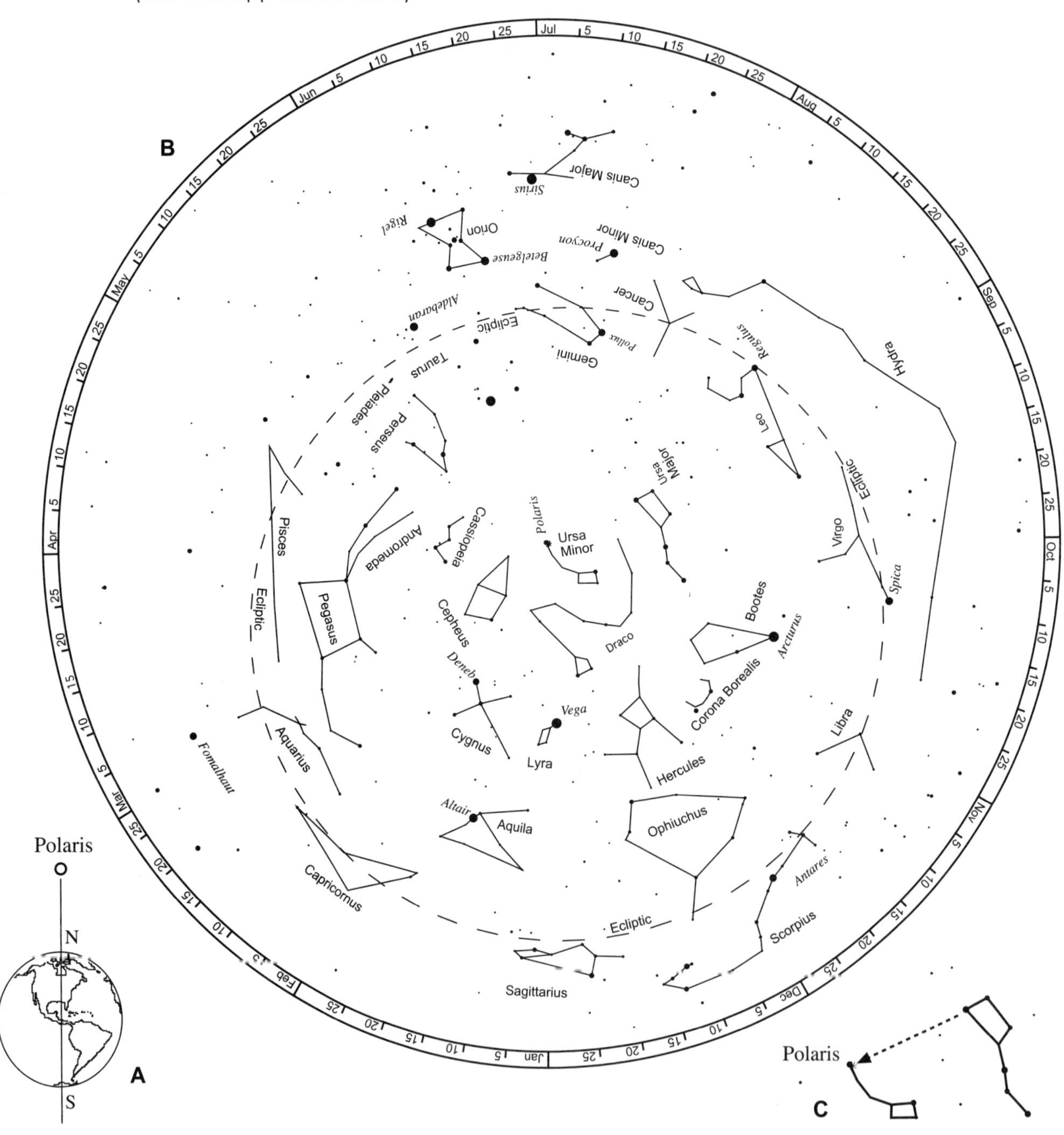

FIGURE 2.12.

STAR WALK APP DETAILS

(*Teacher note: www.apple.com* App Store)

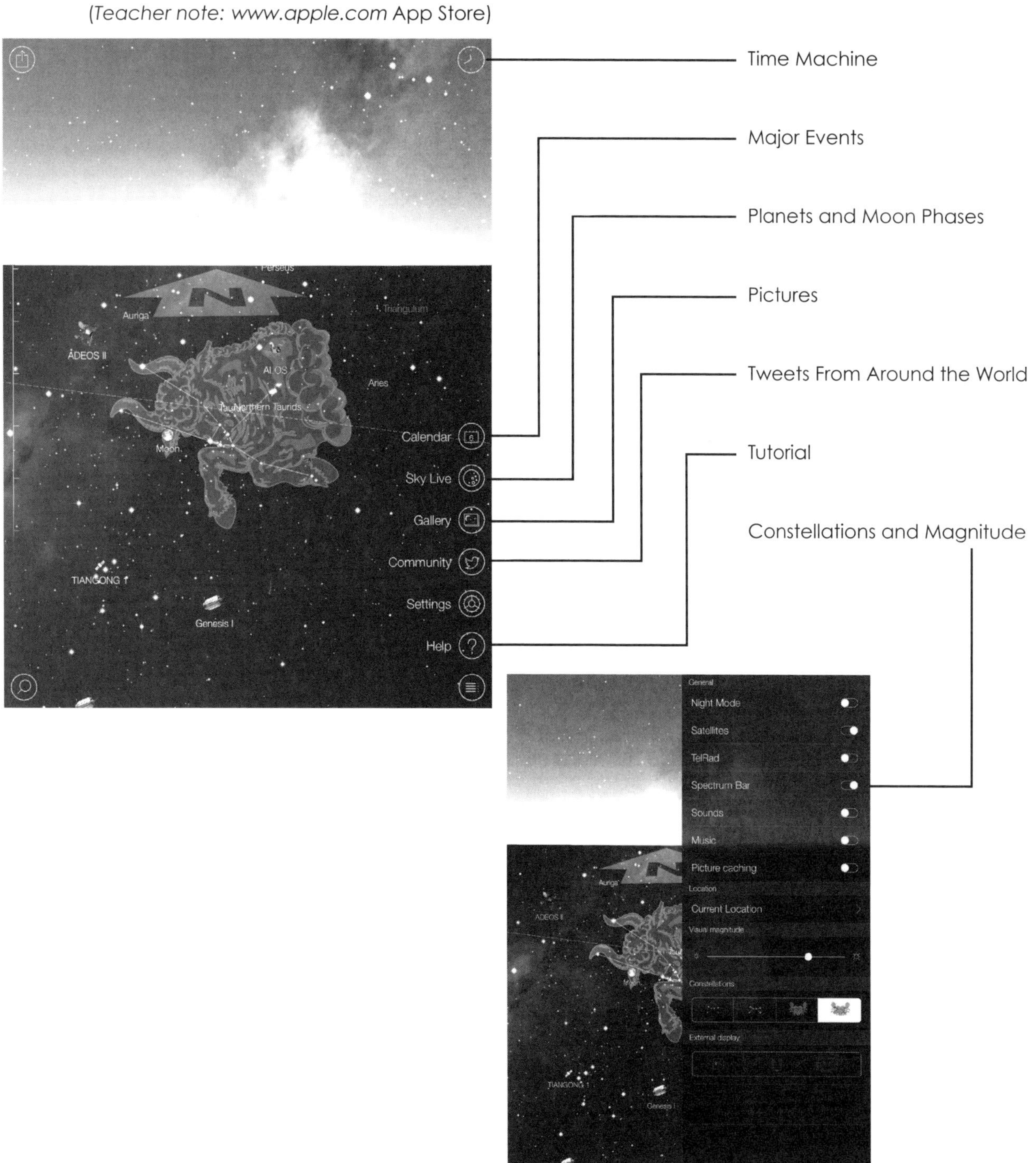

21. Finding That Star Lab

NGSS Alignment

MS-ESS1-2. Develop and use a model to describe the role of gravity in the motions within galaxies and the solar system.

MS-ESS1-3. Analyze and interpret data to determine scale properties of objects in the solar system.

5-ESS1-2. Represent data in graphical displays to reveal patterns of daily changes in length and direction of shadows, day and night, and the seasonal appearance of some stars in the night sky.

2J.

Rafting Through the Constellations Activity

RAFT Story on a Constellation

1. Work with a partner to complete the following writing assignment, based on the **RAFT** (Role, Audience, Format, Topic) method, as described below.
    - **Role:** The role, or point of view, of the story is that of an adolescent. A student in our grade, in our school, is writing your story.
    - **Audience:** The story's audience is our community, and at least three well-known landmarks or important people should be included in your story.
    - **Format:** The format is historical fiction, which means the story must include accurate details about our community, but the author should create the storyline from imagination.
    - **Topic:** The story's topic is how a particular constellation, shining in the night sky above our community, came into being. The character(s) pictured in the constellation should include well-known local landmarks or important people.
2. Read the example below, "The Legend of Orion the Hunter," which involves landmarks and important people in the author's world.
3. Collaborate with your partner to write your own RAFT Story. Be prepared to share your story with classmates.

The Legend of Orion the Hunter

Paul was a brave and wise young boy who lived with his grandfather in the village of Dakota. Paul helped his grandfather around the house, worked in the garden, and took care of the animals on their farm. There was not much time for Paul to play. Each week, Paul walked with his grandfather to Lake Batavia to get water.

One day, his grandfather became very ill. "Paul, you must go for water by yourself," his grandfather told him. "But you must only go to Lake Batavia. Do not go to the Cedar River," he warned.

"But, Grandfather," Paul pleaded. "The water in the Cedar River is very fresh and clear. It twinkles like the stars at night. And it is much closer than Lake Batavia. I can be back with water very fast, if I go to the river."

"That is true, my grandson," he said. "But the water in the Cedar River belongs to Orion, the fierce hunter of the Dakota region. Orion will kill you if you go near his river."

Paul started down the road to Lake Batavia. He had worked very hard that week and quickly became tired. "I can't possibly walk all the way to Lake Batavia," he thought. "I will take only a little water from the Cedar River. Orion the Hunter will never know." So Paul walked across the field to the river.

Just as Paul dipped his bucket into the water, Orion the Hunter burst forth, with a brilliant fire on the tip of his sword, and sparkling gems lining his belt. "What are you doing near my river?" the hunter bellowed fiercely.

Paul was so startled that he almost dropped his bucket in the river. "I am only getting a very small amount of water for my grandfather," he pleaded. "He is very sick, and I am so tired from working all week. I cannot walk all the way to Lake Batavia. Please, you have so much water here in this river. I only need a little."

The hunter paused for a moment. "Well, alright," he mumbled. "Just this once: fill your buckets and leave immediately."

"Thank you, Orion, sir," Paul said. Paul began filling his bucket with the pure river water. When he had filled his buckets, Paul thanked Orion again and left.

The next week, Paul again was going for water. "I should tell Orion the Hunter that my grandfather is feeling better," he thought. So he headed down to the river.

As Paul approached the Cedar River, Orion burst forth, shooting a brilliant fire off the tip of his sword. "What are you doing near my river?" the hunter bellowed.

"It is Paul," he answered quickly. "I came to tell you that my grandfather drank your water and now he is finally feeling better. May I please have a little more water from your river?"

Orion felt sorry for Paul. He muttered, "Well, alright, just this once: fill your buckets and leave."

The next week Paul met a kind-hearted, beautiful young girl, named Lara, with one small water bucket. "Are you going for water with that tiny bucket?" he asked.

"My mother is sick, and this bucket is as large as I can carry," Lara said.

Paul led Lara down to the Cedar River. As they approached, Orion's brilliant belt forewarned of the fire-tipped sword that would soon threaten them. "What are you doing near my river?"

"It is Paul with a girl, named Lara, and she has one very small water bucket. She needs your water to save her mother, who is very sick" he said.

"That is a small bucket," Orion said. He lowered his shining sword and mumbled, "Well, alright, just this once: both of you fill your buckets and leave."

Both Paul and Lara thanked Orion for their water.

Week after week, Paul and Lara went to the Cedar River to fill their buckets. Each time, Orion would burst forth with his brilliant belt and flaming sword and ask, "What are you doing near my river?" Then he would hear their story and say, "Well, alright, just this once: fill your buckets and leave."

One week, when Paul and Lara were going to the river, they met a strong and generous young man, named Daniel. The young man was carrying two buckets.

"Are you going for water?" Paul asked.

"Yes," said Daniel. "But I am afraid I will not be able to walk much farther. I am carrying these two buckets in order to help others out."

"It is not far," replied Paul, "just to the river."

"You cannot get water from the river. Orion the Hunter will hit you with his flaming sword. You must only get water from Lake Batavia," commented Daniel.

"That is silly," said Paul. "Orion the Hunter is my friend. He will not hurt a generous young man like you. Come with us."

Paul did not want Orion to burst forth and scare Daniel, so he ran ahead. "Orion", he called. "I have brought a generous young man carrying two buckets, to help others out. He cannot make it to the lake. Please, let him have some river water. And, please do not threaten him with your fiery sword because that might push him away."

Slowly, Orion came out of the water. He lowered his shining sword and said, "Well, alright, just this once: fill your buckets and leave."

The generous young man thought his new friend Paul was exceedingly clever, and told the village about taking water from the Cedar River. Soon many people came to the river and filled their buckets. Orion never bothered them.

When Orion continued to not harm or threaten them, the people began to tell stories that the fierce hunter must have been killed. Orion seemed to no longer guard the Cedar River.

Paul could not believe these stories. He ran to the river, "Orion, Orion," he shouted. But Orion did not appear. Paul ran up and down the river, searching for Orion. Finally, Paul collapsed, yelling, "I did not mean to hurt you, Orion. You are my friend. Please talk to me," he pleaded.

Very slowly, Orion and his brilliant belt appeared above the water. "Please do not be frustrated," he said to Paul. "I am at fault. I was given this Cedar River water

by the gods of sea and sky to protect. I have failed to protect it. I will never get to protect any water again."

"But you are so kind and good," replied Paul. "I will go to the temple tonight and talk to the gods of sea and sky. There must be other water for you to protect."

That night Paul, Lara, and Daniel, who were now good friends, went to the temple. "Please, give Orion new water. We don't want him to die," they begged. "He is good and has helped many of us," they said. They told the gods of sea and sky many stories of Orion's kindness. The gods were very impressed by Orion.

As the three walked home in the dark, suddenly a bright flash was seen in the sky. They all looked up. There, high above them was the tip of Orion's brilliant shining sword, proudly held to protect all of the waters of the seas and the stars of the sky. Orion the Hunter was positioned directly across from a big water dipper constellation to remind people of his kindness. Paul, Lara, and Daniel knew that Orion the Hunter was finally happy.

2J. Rafting Through the Constellations Activity

NGSS Alignment

5-ESS1-2. Represent data in graphical displays to reveal patterns of daily changes in length and direction of shadows, day and night, and the seasonal appearance of some stars in the night sky.

3
EARTH'S SURFACE
PROCESSES

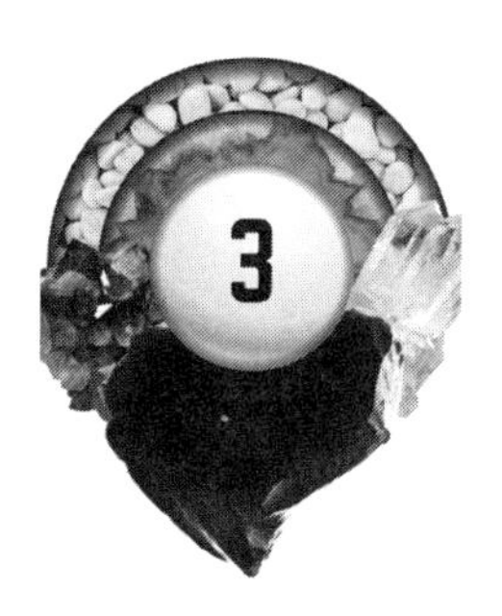

3A.

Periodic Puns Activity

Directions

Elements from the periodic table combine with each other as atoms and molecules to form all of the minerals that we find in our rocks. With some imagination and a pun now and then, it is possible to use the names of elements as synonyms or substitutes for some phrases. So "cesium" your pen and fill in the blanks (Data Table 3.1). But be careful, because spelling counts!

(*Teacher note:* A great NOVA *Elements* app, produced by WGBH Educational Foundation, is now available for both Mac and Windows platforms. It is accessible at *www.pbs.org/wgbh/nova/physics/elements-ipad-app.html*.)

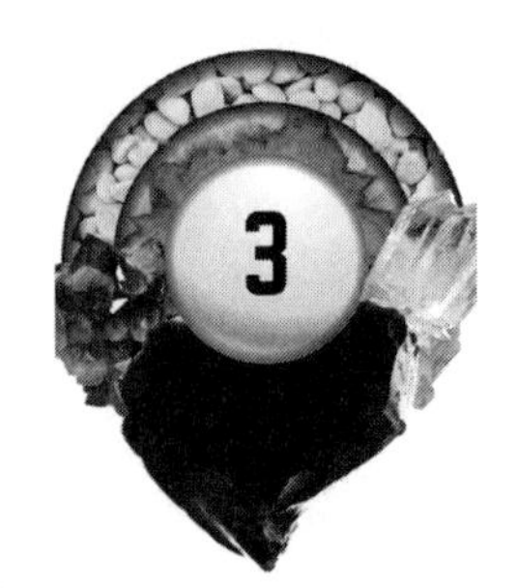

DATA TABLE 3.1.

PERIODIC PUNS

ELEMENT NAME	ELEMENT SYMBOL	ATOMIC NUMBER	PHRASE
1.		#	What a good doctor does for her patients
2.		#	Police officer
3.		#	Have went (very poor grammar)
4.		#	Funeral chant (very rude)
5.		#	Holmium × 0.5 =
6.		#	Chemical Apache (politically incorrect)
7.		#	To press a shirt
8.		#	To guide (past tense)
9.		#	A kitchen work area with a drain
10.		#	An amusing prisoner
11.		#	The Lone Ranger's horse
12.		#	Mickey Mouse's dog
13.		#	It's found on a "sul"
14.		#	View by a person named Cal
15.		#	Two equal a dime
16.		#	Funds from your mother's sister
17.		#	What carpenters do before the roofing
18.		#	Superman's greatest enemy
19.		#	Name of a Nobel Prize winner
20.		#	The planet closest to the Sun
21.		#	You knelt down to put your ___ it
22.		#	The thing that grinds garbage

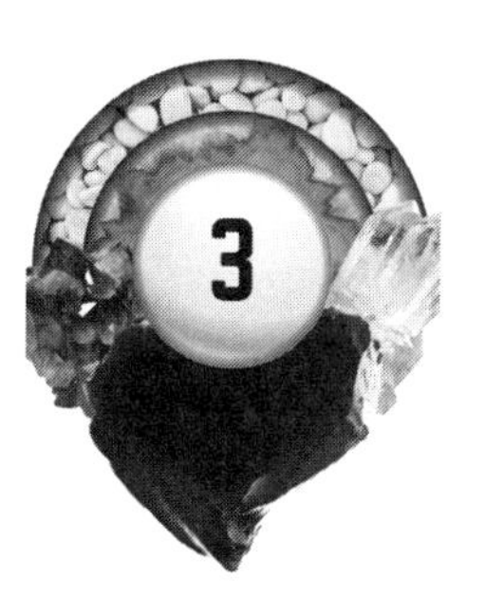

ANSWERS FOR DATA TABLE 3.1.

1. Helium He #2 and Curium Cm #96
2. Copper Cu #29
3. Argon Ar #18
4. Barium Ba #56
5. Hafnium Hf #72
6. Indium In #49
7. Iron Fe #26
8. Lead Pb #82
9. Zinc Zn #30
10. Silicon Si #14
11. Silver Ag #47
12. Plutonium Pu #94
13. Sulfur S #16
14. Calcium Ca #20
15. Nickel Ni #28
16. Antimony Sb #51
17. Fluorine F #9
18. Krypton Kr #36
19. Einsteinium Es #99
20. Mercury Hg #80
21. Neon Ne #10
22. Disprosium Dy #66

3A. Periodic Puns Activity

NGSS Alignment

MS-ESS2-1. Develop a model to describe the cycling of Earth's materials and the flow of energy that drives this process.

MS-PS1-1. Develop models to describe the atomic composition of simple molecules and extended structures.

5-PS1-1. Develop a model to describe that matter is made of particles too small to be seen.

5-PS1-3. Make observations and measurements to identify materials based on their properties.

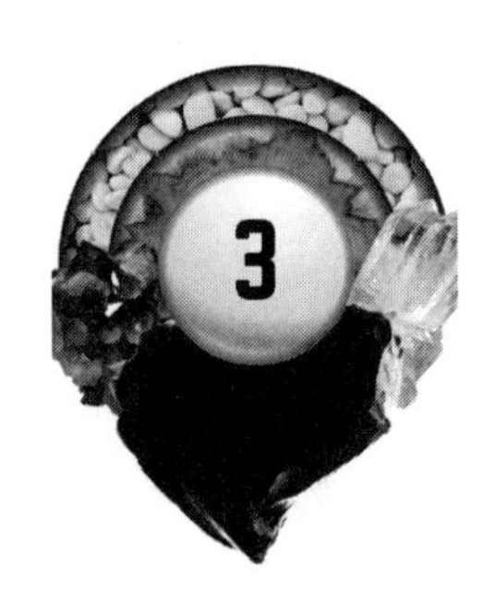

3B.

Weighing in on Minerals Lab

Problem

What methods can geologists use to identify different minerals?

Prediction

Describe how knowing the density of a stone can help you identify it.

(*Teacher note:* Have students share several predictions out loud, so misconceptions can be anticipated and explained.)

Thinking About the Problem

How can scientists figure out which mineral they are examining? Minerals are Earth materials that have four main characteristics: they are solid, inorganic (not a mixture of carbon, hydrogen, oxygen), naturally occurring, and have a definite chemical structure. Minerals are identifiable based on a number of different properties. One is the mineral's particular *crystal structure*. For example, quartz has a hexagonal (*hexa* "six" in Greek) crystal form. Another is its planes, or *cleave lines*, along which a mineral is weakest and tends to fracture easily. Another is the *luster*, or the way that light tends to reflect off of a mineral's surface. Still another is the color left behind, called *streak*, when a mineral scratches a porcelain surface. Scientists use these methods to identify minerals. In geology (*geo* "earth" in Greek), you will learn about two diagnostic tests for minerals: density and hardness.

How can you tell how dense rocks and minerals are? Geologists have developed tools that help to measure density. Either a triple beam balance or an electronic balance can be used for this purpose. In addition, if the mineral to be measured is a regularly shaped object, such as a cube, its volume can be determined with a mathematical formula: Length × Width × Height.

To measure the volume of objects that do not have a regular shape, the mathematical formula given above for volume does not work. In those cases, a water displacement method can be used. Density (comparable to "specific gravity," which is often used in geology) determines the mass of the mineral in relation to the mass of

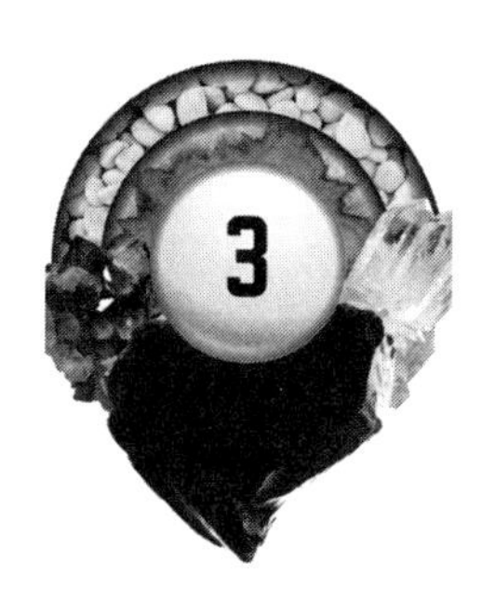

an equal volume of water. Mineral densities can range from 1 to 20. For example, since quartz has a density of 2.65, this means that it is 2.65 times as heavy as the same volume of water. Density values under 2 for minerals are regarded as *light* (amber is 1.0). *Normal* is the term given to minerals with density values from 2 to 3 (calcite is 2.7). All minerals above 3.0 are considered *heavy* (galena is 7.4). Testing for density is how people determined if they had found gold during the Gold Rush in early American history.

Write three main points from the "Thinking About the Problem" reading:

1. A mineral is ...

2.

3.

Procedure

1. Draw a labeled sketch of your experimental materials in the box on p. 116.
2. Using a triple beam balance, determine and record the mass of all six minerals in Data Table 3.2 (p. 117).
3. Using the water displacement method with the graduated cylinder, determine and record the volume of all six minerals in Data Table 3.2 (p. 117).
4. Calculate the density of each object. ($D = M \div V$) and record in Data Table 3.2 (p. 117).
5. Record your density results in the class average data table (Data Table 3.3, p. 118).
6. Check the minerals. Don't wreck the minerals.

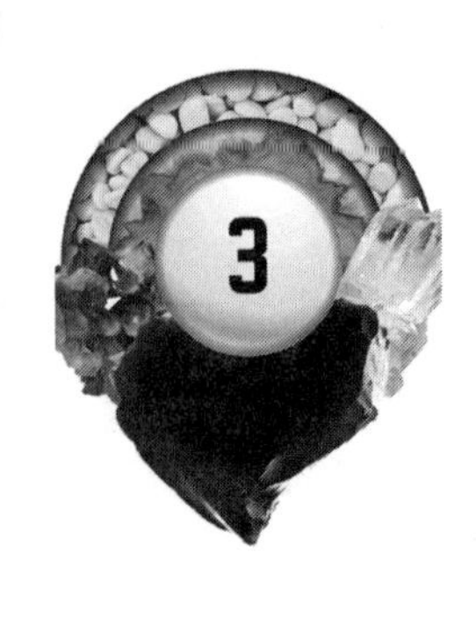

DRAW LABELED SKETCH OF EXPERIMENTAL MATERIALS HERE.

Analysis

1. If you were given a ring, how could you determine if it was pure gold?
2. If gypsum has a known density of 2.3, which of your minerals is most likely gypsum?
3. If halite has a known density of 2.2, which of your minerals is most likely halite?
4. If quartz has a known density of 2.6, which of your minerals is most likely quartz?
5. If calcite has a known density of 2.7, which of your minerals is most likely calcite?
6. If magnetite has a known density of 5.0, which of your minerals is most likely magnetite?
7. If feldspar has a known density of 2.55, which of your minerals is most likely feldspar?
8. (Enrichment) Explain a method that you could use to determine your own body's density.
9. (Enrichment) Determine the density of two regularly shaped objects (wooden block, dice, for example) using both of the volume determination methods. Account for any differences you find.

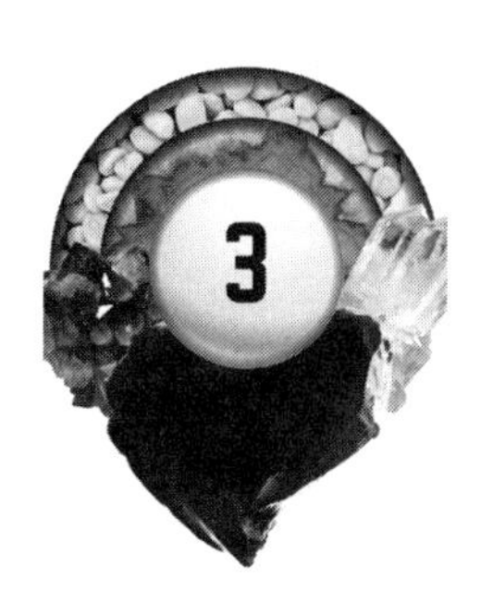

10. (Enrichment) Design and plan an experiment that could be used to determine the year in which the composition of pennies was changed from copper to zinc.

Learning Target

Students use density measurement methods to identify minerals.

I Learned:

Redo:

Manipulated Variable:

Measured Variable:

Controlled Variable:

DATA TABLE 3.2.

DENSITY OF MINERALS (SMALL GROUP)

MINERAL	MASS (G)	VOLUME (ML)	DENSITY (G/ML)
1			
2			
3			
4			
5			
6			

(*Teacher note:* Mineral samples are: #1 calcite 2.7, #2 halite 2.2, #3 gypsum 2.3, #4 feldspar 2.55, #5 quartz 2.6, and #6 magnetite 5.0.)

DATA TABLE 3.3.

DENSITY OF MINERALS (CLASS AVERAGE)

	GROUP RESULTS ON DENSITY OF EACH MINERAL (G/ML)							
MINERAL	GROUP A RESULTS	GROUP B RESULTS	GROUP C RESULTS	GROUP D RESULTS	GROUP E RESULTS	GROUP F RESULTS	GROUP G RESULTS	AVERAGE DENSITY (G/ML)
1								
2								
3								
4								
5								
6								

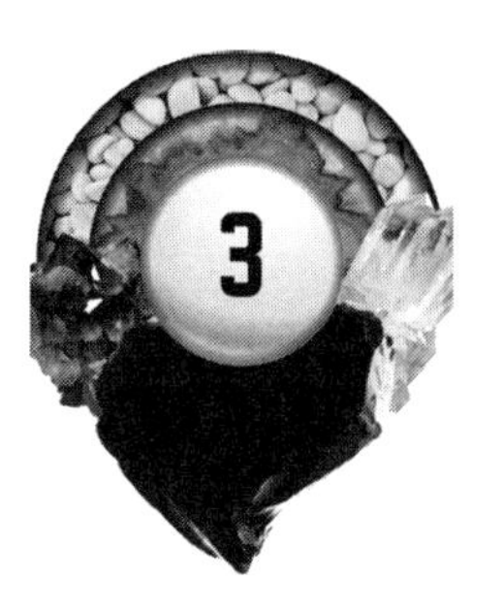

Density Concept Flow Map

A concept flow map (such as Figure 3.1) can be used to develop a better understanding of the various methods used to determine the density of an object. The three examples (rectangular objects, smaller nonrectangular objects, and larger nonrectangular objects) are represented in the figure below, which was modeled on a student's notebook entry.

FIGURE 3.1.

CONCEPT MAP ON DENSITY

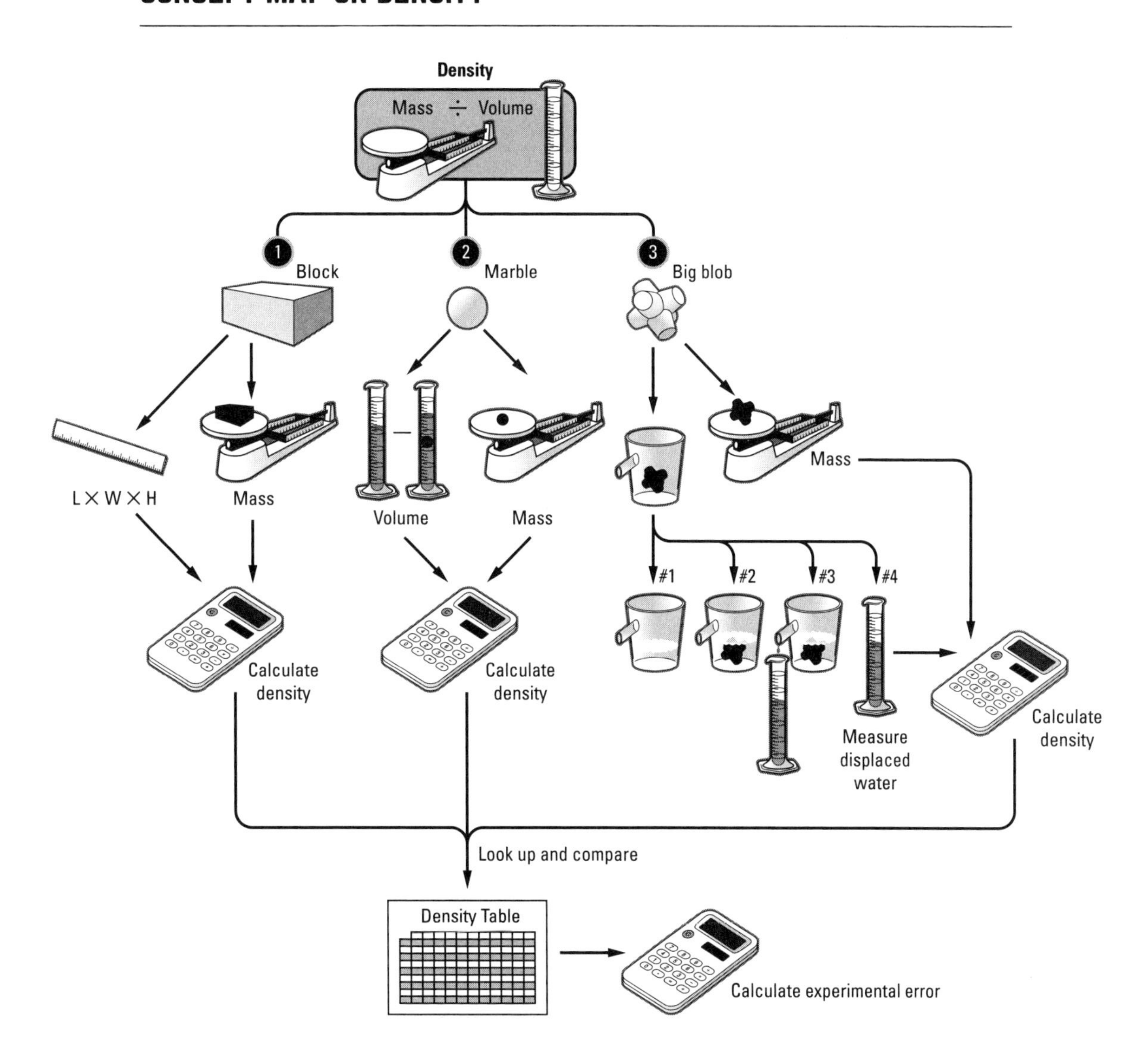

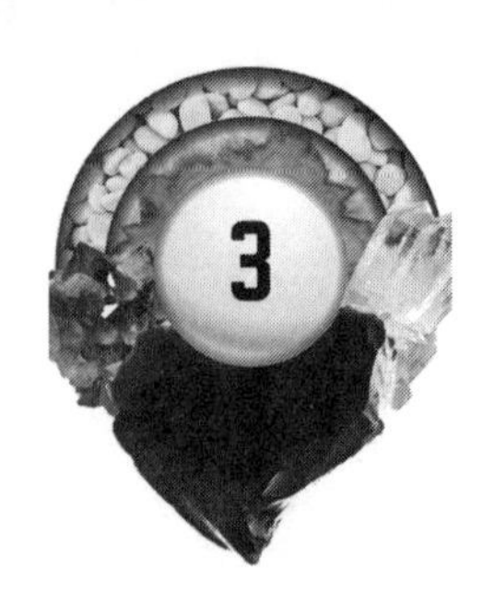

3B. Weighing in on Minerals Lab

NGSS Alignment
MS-ESS2-1. Develop a model to describe the cycling of Earth's materials and the flow of energy that drives this process.

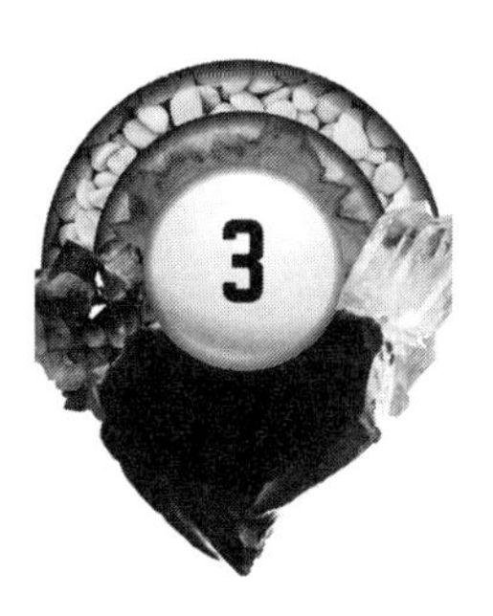

3C.

Knowing Mohs Lab

Problem

Why are diamonds considered to be the hardest known mineral?

Prediction

Give a working definition of *mineral*.

(*Teacher note:* Have students share several predictions out loud, so misconceptions can be anticipated and explained.)

Vocabulary for Glossary

- *Periodic table of elements:* List of all the atoms that are fundamental building blocks for all things on Earth and in space. Elements make up minerals.
- *Mineral:* Naturally occurring elements, which exist in a crystal form (patterns). Minerals make up rocks and gemstones.
- *Rock:* Natural mixture of minerals. They can be formed into one of three groups: igneous, metamorphic, and sedimentary.

Thinking About the Problem

Do you know why diamonds are often used as industrial tools? The answer is because they are so hard they can be used to cut metals and other hard objects. In fact, the gemstone diamond is the hardest known mineral. Do you know why talcum powder is so soft? This powder comes from pulverizing the mineral talc, which is very soft. For geologists, the hardness of a mineral is determined by how easily it can be scratched compared with other minerals on the Mohs hardness scale (named after German mineralogist Frederich Mohs).

Scientists use various properties—such as hardness, luster, color, and specific gravity (density)—to help identify minerals. In this lab, you will learn to identify a set of minerals by comparing their hardness values against known standards. Geologists frequently use glass (hardness value of 5.5), copper (hardness of 3.5), and a fingernail (hardness of 2.5) as standards.

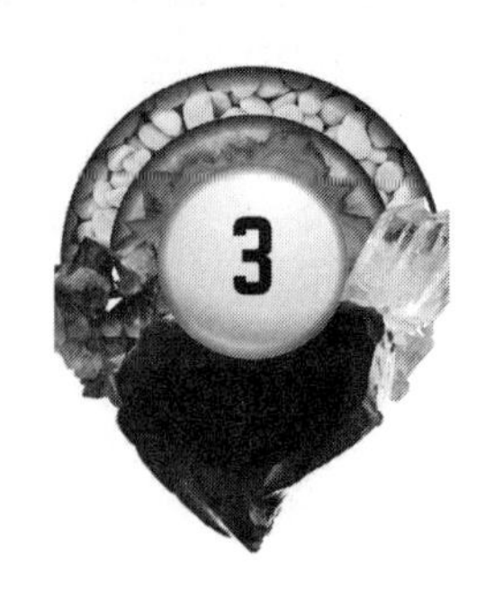

Write three main points from the "Thinking About the Problem" reading:

1. Hardness of a mineral is determined by ...

2.

3.

Procedure

1. Draw a labeled sketch of your experimental materials in the box below
2. Test the hardness of each mineral by using the tools at your lab station. Record your results as *yes* or *no* in Data Table 3.4.
3. Use Data Table 3.5 and Data Table 3.6 (p. 124) to help you estimate the Mohs hardness value based upon your experiment results. Record your answer in each case as *less than* (<) or *greater than* (>).
4. Check the mineral. Don't wreck the mineral.

DRAW LABELED SKETCH OF EXPERIMENTAL MATERIALS HERE.

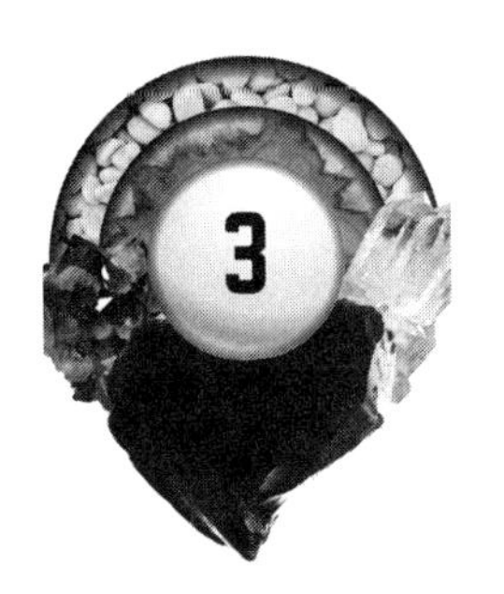

DATA TABLE 3.4.

HARDNESS TEST RESULTS

MINERAL SAMPLE #	FINGERNAIL TEST	COPPER PLATE TEST	IRON NAIL TEST	GLASS PLATE TEST	HARDNESS VALUE	MINERAL NAME
1						
2						
3						
4						
5						
6						

DATA TABLE 3.5.

MOHS HARDNESS SCALE

TEST RESULT	HARDNESS ESTIMATE	
	YES	NO
Can it be scratched by your fingernail?	< 2	> 2
Can it be scratched by the copper plate?	< 3	> 3
Can it be scratched by the iron nail?	< 5	> 5
Can it scratch the glass plate?	> 6	< 6

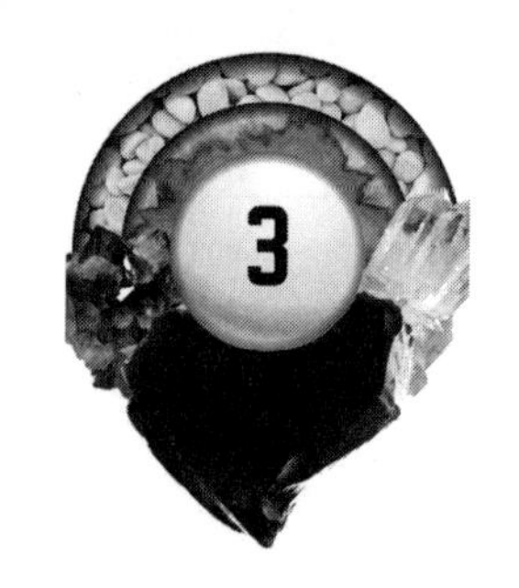

DATA TABLE 3.6.

MOHS MINERAL HARDNESS VALUES

HARDNESS	MINERAL	SCRATCH TEST RESULTS
1.0	Talc	Easily scratched by fingernail
1.5	Graphite	Easily scratched by fingernail
2	Gypsum	Scratched by fingernail
2.25	Galena	Scratched by fingernail
2.25	Halite	Scratched by fingernail
2.5	Muscovite	Scratched by fingernail
2.5	Cinnabar	Scratched by fingernail
2.75	Biotite	Not quite scratched by copper plate
3	Calcite	Slightly scratched by copper plate
4	Fluorite	Easily scratched by iron nail
5	Apatite	Scratched by iron nail
5.5	Pyroxine	Scratched by iron nail
5.75	Magnetite	Slightly scratched by iron nail; it might scratch glass plate
6	Feldspar	Slightly scratched by iron nail; it scratches glass plate
6.25	Pyrite	Barely scratched by iron nail; it scratches glass plate
6.5	Olivine	Scratches glass plate
7	Quartz	It scratches both iron nail and glass plate
8	Topaz	It scratches quartz
9	Corundum	It scratches topaz
10	Diamond	Hardest known mineral

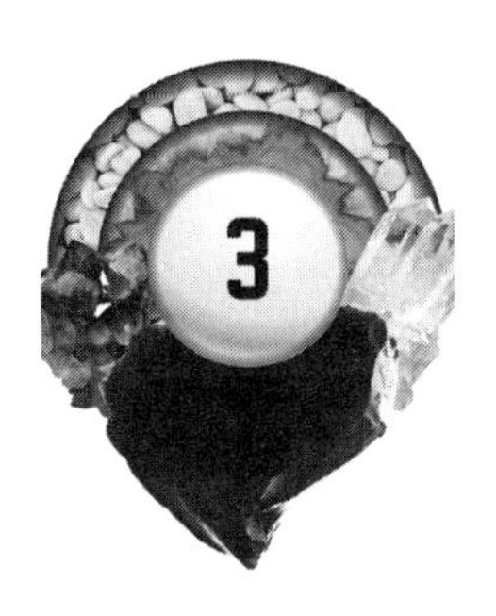

Analysis

1. Which method should you use to identify a mineral most accurately: its hardness value, appearance, or density? Explain.
2. Imagine you have a mineral with a Mohs scale of 2. Describe the procedure for determining this hardness value.
3. Imagine you have a mineral with a Mohs scale of 6. Describe the procedure for determining this hardness value.
4. Describe the relationship between a mineral's hardness and its ability to be worn away by the processes of either weathering or erosion.
5. Use the Mohs Hardness Scale posted to name each of your unknown minerals in Data Table 3.4. (*Teacher note:* Minerals are as follows: #1 Calcite 3, #2 Halite 2.25, #3 Gypsum 2, #4 Feldspar 6, #5 Quartz 7, and #6 Magnetite 5.75.)
6. (Enrichment) Research and then list one functional use of each mineral studied in this lab.

Learning Target

Identify a set of minerals by comparing their hardness values against known standards.

I Learned:

Redo:

Manipulated Variable:

Measured Variable:

Controlled Variable:

3C. Knowing Mohs Lab

NGSS **Alignment**

MS-ESS2-1. Develop a model to describe the cycling of Earth's materials and the flow of energy that drives this process.

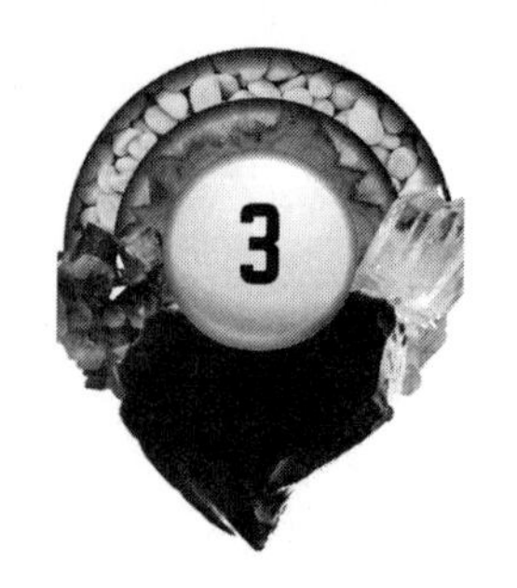

3D.

Classifying Rocks and Geologic Role Lab

Problem

What are the differences between igneous, sedimentary, and metamorphic rocks?

Prediction

Give a working definition of *rock*.

(*Teacher note:* Have students share several predictions out loud, so misconceptions can be anticipated and explained.)

Thinking About the Problem

What are rocks made of? A rock is a natural mixture of minerals. Of the 3,000 known minerals, only a few dozen are essential constituents of rocks. A "felsic" mineral, which is a combination of the terms feldspar and silica, is among the most common minerals that make up rocks. Examples of felsic minerals are quartz, feldspar, and mica. The other most common mineral that makes up rocks is a "mafic" mineral, which is a combination of the terms that describe magnesium and iron. Examples of mafic minerals are olivine, pyroxene, and amphibole.

The rock cycle (Figure 3.2, p. 129) is the direct result of energy flowing and matter cycling within and throughout the Earth's systems. There are three main types of rocks: igneous, sedimentary, and metamorphic. These are classified by the role that the Earth's energy played in their development. Igneous (*igneus* "fire" in Latin) rocks develop when liquid molten rock, called magma, solidifies in the Earth's crust or on the Earth's surface. The rate at which the magma cools plays a big role in the crystal size and mineral composition.

Sedimentary (*sedere* "to settle" in Latin) rocks develop at the Earth's surface as the weathering result of rocks exposed to wind, water, or ice. When weather and other forces of erosion wear away rocks, sediments form. Those sediments can be

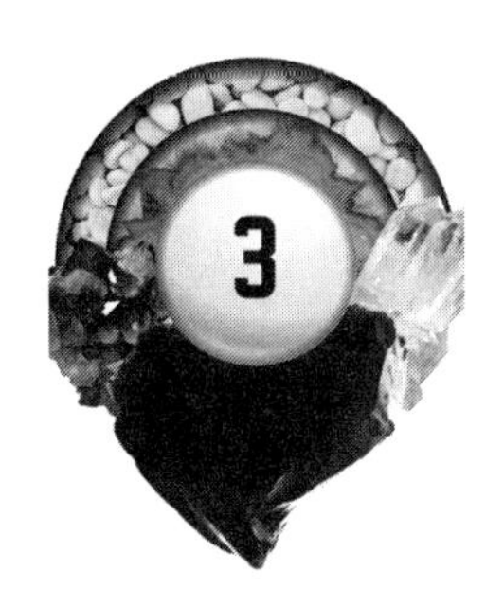

DRAW LABELED SKETCH OF EXPERIMENTAL MATERIALS HERE.

compacted, through the process of lithification (*lithos* "stone" in Greek), to form sedimentary rocks.

Metamorphic (*metamorphoun* "to change" in Greek) rocks develop through the transformation of other rocks due to great pressures or high temperatures. Those tremendously strong forces can change preexisting rocks through the process of metamorphism. Rocks remain solid during the entire process.

The rock cycle describes the roles and relationships between all three of these rock groups. In this lab, you will use a rock classification system, called the Rock and Role Classification Key (Figure 3.3, p. 130), to determine into which group several unknown rocks fit. Draw a sketch of your materials in the box above.

Write three main points from the "Thinking About the Problem" reading:

1. Igneous rocks develop …

2. Sedimentary rocks develop …

3. Metamorphic rocks develop …

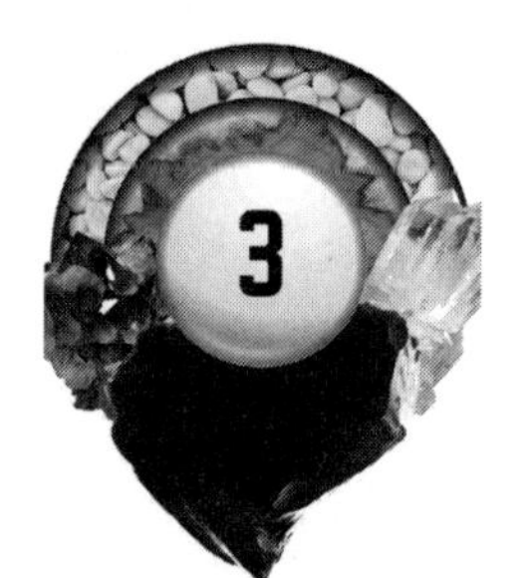

DATA TABLE 3.7.

CLASSIFICATION OF ROCKS AND GEOLOGIC ROLE

ROCK SAMPLE #	DESCRIPTION OF VISIBLE PROPERTIES	ROCK CLASSIFICATION
1		
2		
3		
4		
5		
6		
7		
8		
9		

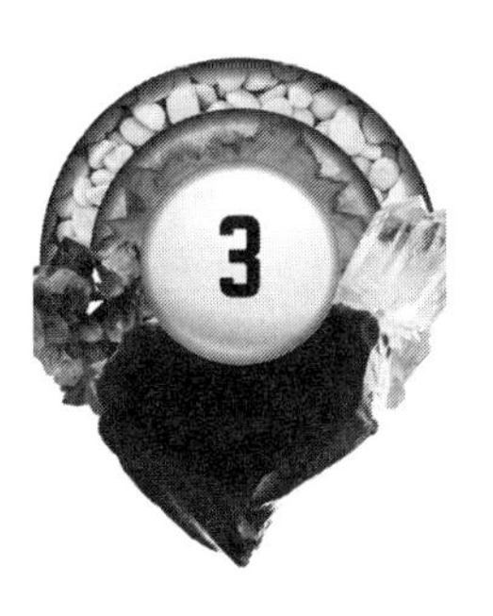

FIGURE 3.2.

THE ROCK CYCLE DIAGRAM

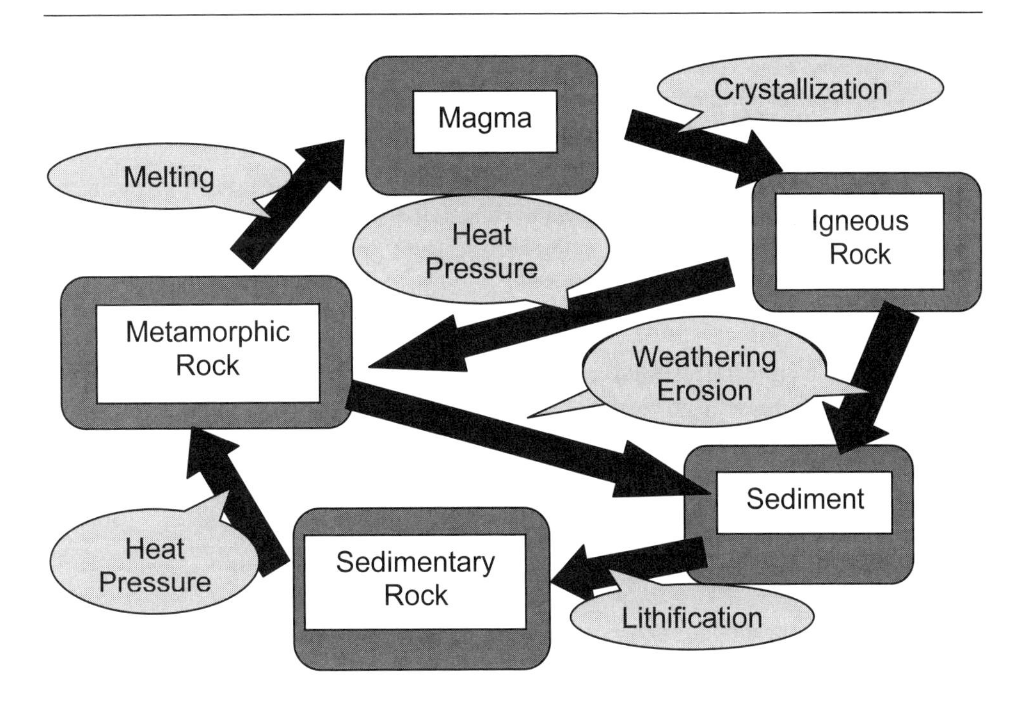

Learning Target

Use a rock classification system, called the Rock and Role Classification Key (Figure 3.3, p. 130), to determine into which group several unknown rocks fit (Data Table 3.7).

I Learned:

Redo:

Manipulated Variable:

Measured Variable:

Controlled Variable:

Teacher note: Rock samples are as follows: three igneous (obsidian #1, basalt #9, pumice #5), three sedimentary (sandstone #2, shale #8, conglomerate #4), and three metamorphic (quartzite #3, slate #7, marble #6).

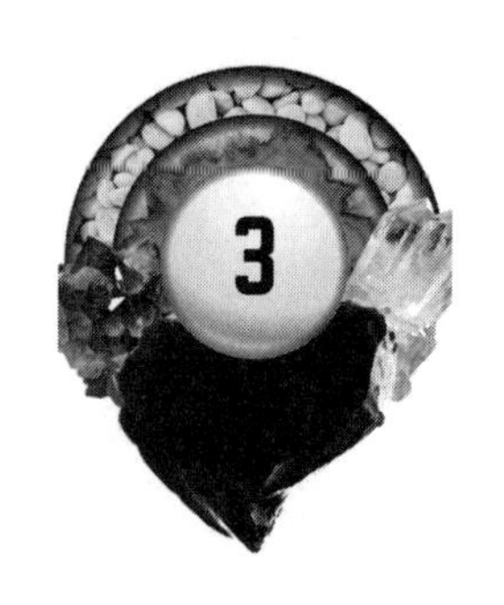

FIGURE 3.3.

ROCK AND ROLE CLASSIFICATION KEY

ROCK AND ROLE KEY

Directions: Answer each question in sequence. Observe your rock sample carefully. Check the rock sample. Don't wreck the rock sample.

1. **Is the rock made up of easily visible, separate particles of crystals or sand?**

   A. Yes, particles are easily visible. [Go to step 3]

   B. No, it is not composed of easily visible, separate particles. [Go to step 2]

2. **Is the rock completely solid, but not glassy?**

   A. Yes, the rock is completely solid. [Go to step 5]

   B. No, the rock is glassy, or spongelike, or porous. [The rock is igneous.]

3. **What types of particles make up your rock?**

   A. The rock is made of easily visible, shiny mineral crystals. [Go to step 4]

   B. The rock is made up of sand or pebbles that appear to be cemented together. [The rock is sedimentary.]

4. **Do the mineral crystals of your rock tend to line up, or form different-colored bands?**

   A. Yes, the mineral crystals tend to line up or form bands. [The rock is metamorphic.]

   B. No, the mineral crystals are not lined up in a particular direction. [Go to step 7]

5. **Is your rock made up of visible layers, or does it tend to break into layers?**

   A. Yes, the rock has visible layers or tends to break into layers. [Go to step 6]

   B. No, the rock has no layers nor does it break into layers. [The rock is igneous.]

6. **Is your rock relatively reflective or shiny?**

   A. Yes, the rock is fairly shiny or reflective. [The rock is metamorphic.]

   B. No, the rock is quite dull, not reflective or shiny. [The rock is sedimentary.]

7. **Can you see that your rock is made up of one or more different types or colors of mineral crystals?**

   A. Yes, there appears to be two or more types of mineral crystals. [The rock is igneous.]

   B. No, all of the crystals appear to be the same mineral. [The rock is metamorphic.]

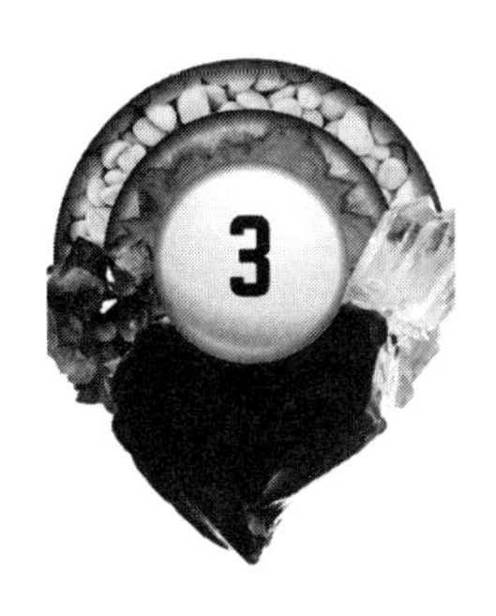

Analysis

1. Give two similarities and two differences between the igneous and metamorphic rocks.
2. Give two similarities and two differences between the sedimentary and metamorphic rocks.
3. Give two similarities and two differences between the igneous and sedimentary rocks.
4. (Enrichment) Specifically, how would you improve the Rock and Role Key, to get rid of any challenges that it presented to you?
5. (Enrichment) Closely examine several sedimentary rocks. In addition to sight, what other senses could help you key out the rocks?

3D. Classifying Rocks and Geologic Role Lab

***NGSS* Alignment**

MS-ESS2-1. Develop a model to describe the cycling of Earth's materials and the flow of energy that drives this process.

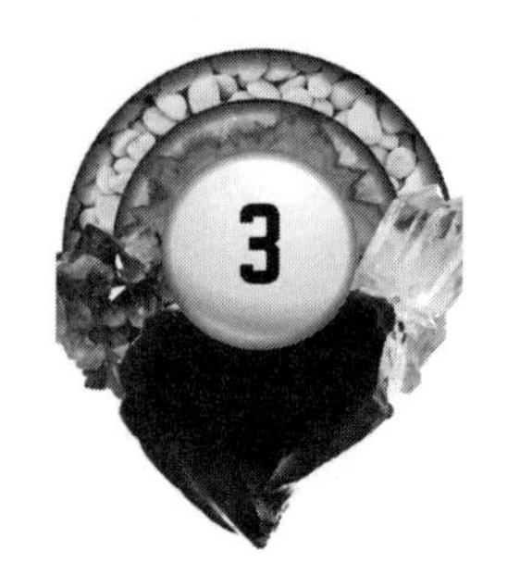

3E.

Edible Stalactites and Stalagmites Lab

Problem

Have you ever imagined eating stalactites or stalagmites? Stalactites grow from the ceilings of caves, gradually getting longer as calcium carbonate ($CaCO_3$) deposits drip down and crystallize. Stalagmites pile up on the floor of caves, beneath the stalactites. You can grow crystals just like them—in your own kitchen. Rock candy (large sugar crystals) grows in a similar way to stalactites and stalagmites.

Prediction

How long do you predict it will take to grow an edible crystal?

(*Teacher note:* Have students share several predictions out loud, so misconceptions can be anticipated and explained.)

Materials

- 250 ml (1 cup) of water
- 500 ml (2 cups) of granulated sugar
- Coffee filter strip (or cotton string)
- Large drinking glass or other glass jar
- Pencil
- Small saucepan

Procedure

1. Clean all materials (saucepan, glass, paper clip) thoroughly—otherwise the crystals will not form.

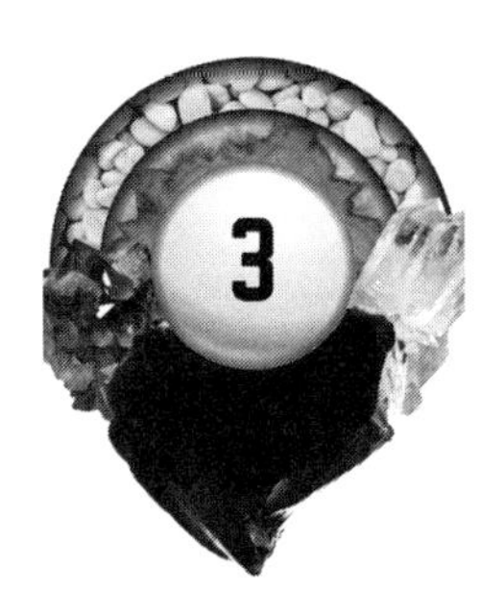

2. Bring the water to a boil in the saucepan.
3. As soon as the water boils, turn off the heat and begin stirring in the sugar.
4. Continue adding the sugar until no more will dissolve.
5. Allow the solution to cool for about five minutes.
6. Pour the solution into the drinking glass.
7. Rub some sugar onto the string so that the crystals will stick to it.
8. Tie one end of the string to the pencil, and then rest the pencil on the rim of the glass (Figure 3.4).
9. Put the glass in a place where it will remain relatively cool and completely undisturbed for several days.

FIGURE 3.4.

STALACTITE AND STALAGMITE SETUP

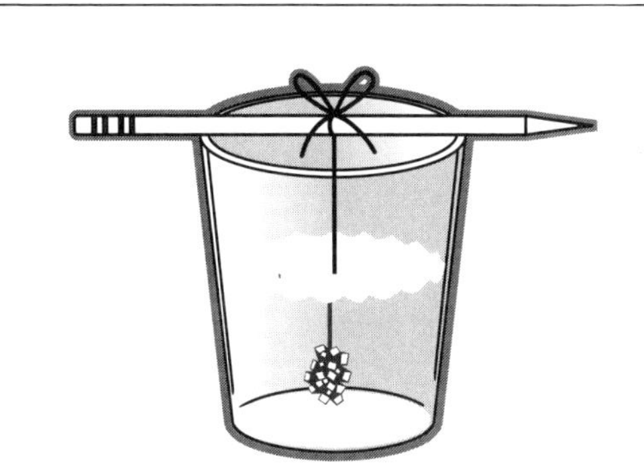

DRAW LABELED SKETCH OF EXPERIMENTAL MATERIALS HERE.

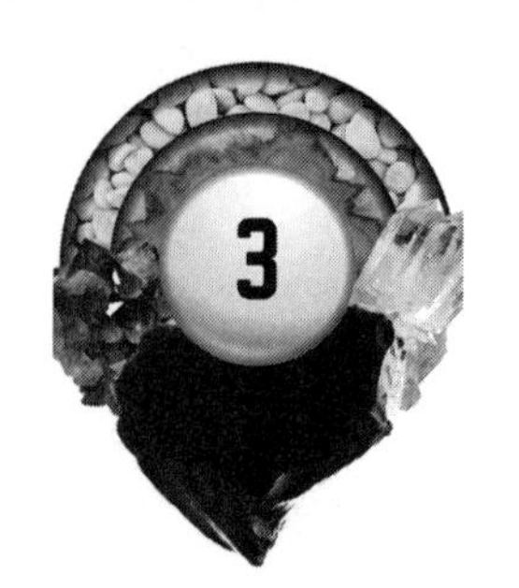

Analysis

Bring your edible stalagmite or stalactite (in a small plastic bag or envelope) in to show your teacher, along with this completed lab report.

3E. Edible Stalactites and Stalagmites Lab

NGSS Alignment
MS-ESS2-1. Develop a model to describe the cycling of Earth's materials and the flow of energy that drives this process.

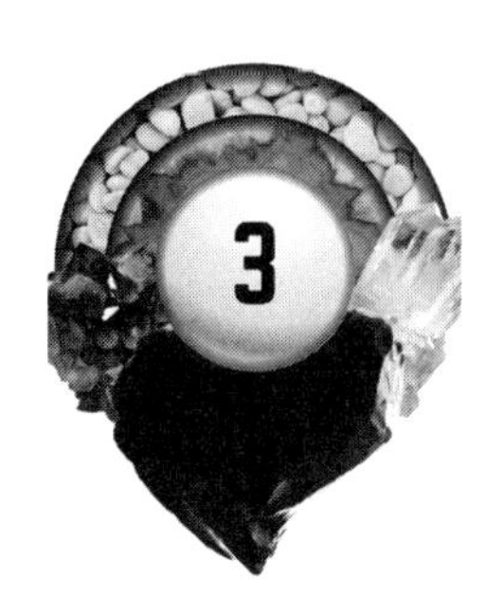

3F.

Weathering the Rocks Lab

Problem

Why is it important for people to know how rocks can be changed by physical and chemical weathering?

Prediction

Give a working definition of *erosion*.

(*Teacher note:* Have students share several predictions out loud, so misconceptions can be anticipated and explained.)

Thinking About the Problem

Weathering is the process by which materials on Earth's surface are broken down and changed in form, either mechanical (physical) or chemical. In mechanical weathering, a material, such as a rock, is changed in size and shape. In chemical weathering, the substances making up matter are changed to other substances.

Mechanical weathering is caused by changes in temperature, pressure, and the actions of plant roots or animals. Frost wedging in cold climate winters frequently causes potholes and broken sidewalks.

Chemical weathering occurs in the presence of water or moisture in the air. Frequently water drips through soils, which overlie the rock layers, getting more acidic as it passes through. When this acidic water hits the rocks, which are made of calcium carbonates ($CaCO_3$), these rocks break down and erode away.

Many examples of physical and chemical changes can be found in your home and neighborhood. Weathering can corrode metals as well as break down rocks. Erosion is what happens after the weathering. When parts of the broken rocks are carried off by river water, for example, then erosion has happened.

Write three main points from the "Thinking About the Problem" reading:

1. Mechanical weathering is …

2. Chemical weathering is …

3. Erosion is …

FIGURE 3.5.

PHOTO SCALE TOKEN

Earth Science Rocks!

Materials

- iPad Camera

Procedure

1. Carefully observe the outdoor world around you, finding examples of weathering (both mechanical and chemical) and of erosion. Photograph each, by placing your token (Figure 3.5) next to each item (the token helps to show scale), documenting details about each as you find them (Data Table 3.8).
2. You have one week to bring in evidence of one item that was chemically weathered, one item that was mechanically weathered, one erosion example, and one that shows an example of how humans are trying to prevent erosion.
3. You may need to use Safari for further ideas in some of these categories, but your photos must be taken by you, and include your Photo Scale Token in them.

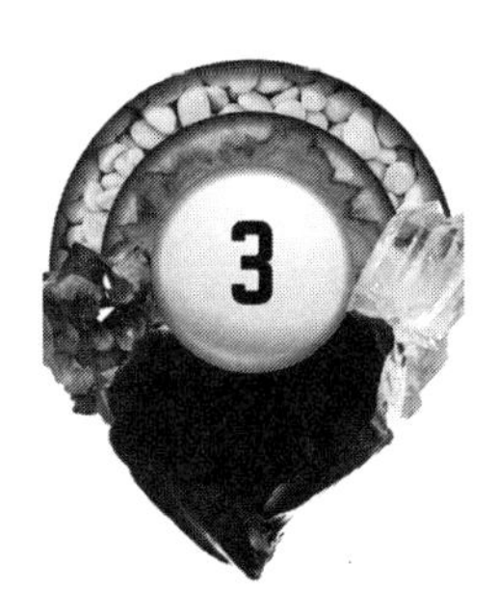

DATA TABLE 3.8 ______________________

(Teacher note: Student provides a descriptive data table title.)

PHOTO TYPE	LABELED IMAGE OF ITEM	DETAILED DESCRIPTION OF LOCATION FOUND	YOUR JUSTIFICATION WHY THIS IMAGE FITS THE REQUIREMENT
Mechanical weathering			
Chemical weathering			
Erosion			
Erosion prevention			

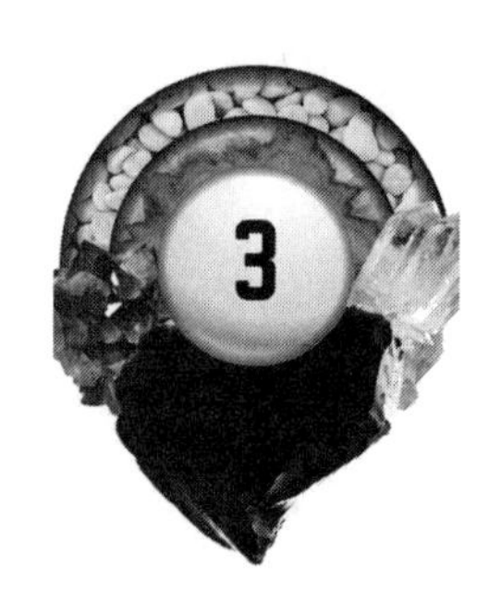

Learning Target

Locating examples of weathering and erosion found in students' neighborhoods.

Analysis

1. What effect do you think acid rain has on limestone building stone blocks?
2. Explain how large temperature changes involving freezing and thawing can help break apart rocks.
3. Give an example of why it is important for people to know how rocks are changed by mechanical and chemical weathering.
4. (Enrichment) Construct an explanation, based on evidence that you gathered in this lab, for how a surface is changed by weathering and/or erosion.

3F. Weathering the Rocks Lab

NGSS Alignment

MS-ESS1-4. Construct a scientific explanation based on evidence from rock strata for how the geologic time scale is used to organize Earth's 4.6-billion-year-old history.

MS-ESS2-1. Develop a model to describe the cycling of Earth's materials and the flow of energy that drives this process.

MS-ESS2-2. Construct an explanation based on evidence for how geoscience processes have changed Earth's surface at varying time and spatial scales.

4-ESS2-1. Make observations and/or measurements to provide evidence of the effects of weathering or the rate of erosion by water, ice, wind, or vegetation.

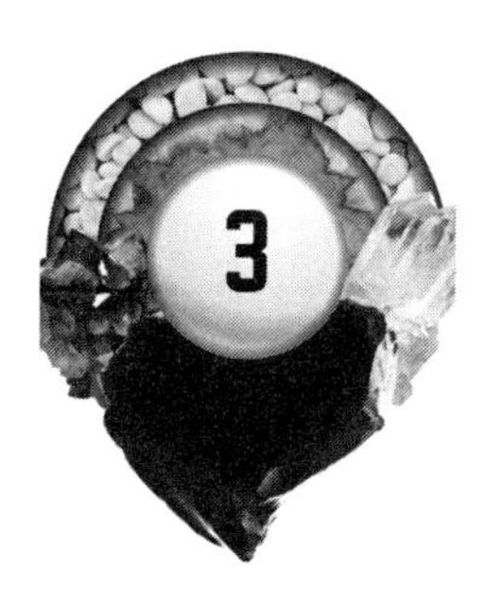

3G.

Hunting Through the Sand Lab

Problem

Can you identify the rocks and minerals in sand?

Prediction

Describe any materials that can be found in sand.

(*Teacher note:* Have students share several predictions out loud, so misconceptions can be anticipated and explained.)

Thinking About the Problem

Why do flowers grow well in one soil but not in another? Soils are typically composed of 45% minerals, 25% air, 25% water, and 5% organic matter. In general, the more organic matter, the better the flowers grow, but proper amounts of air, water, and trace minerals are important, too.

Soil textures are determined by the percentages of clay, silt, and sand in the soil. Like rocks, soils are also found in layers. A description of this layering is called a *soil profile.*

As you go deeper in the soil, you find lesser amounts of partly decayed organic matter and greater amounts of partially weathered bedrock material. All rocks exposed at Earth's surface are subject to weathering, which leads to the formation of sediments. Forces of erosion (primarily water and wind) can transport sediments around Earth's surface. Erosion and weathering form soil.

Drilling into the Earth, scientists do not go very deep relative to all of the layers of the Earth, digging down into the crust only. The surface layers of the Earth are where sand is found. Earthquakes can go very deep into the crustal plates, but still do not go below the thickness of the line shown in Figure 3.6 (p. 140). This structure of the Earth diagram was determined primarily through validations of evidence included in the theory of plate tectonics. In this lab you will examine the various materials that make up the sand that is found in a soil sample.

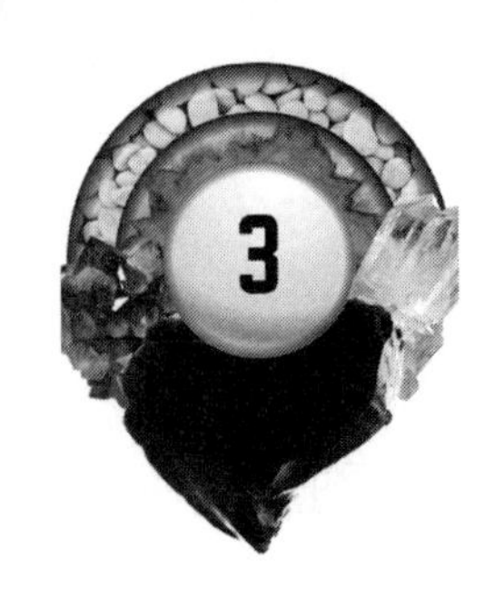

FIGURE 3.6.

STRUCTURE OF THE EARTH

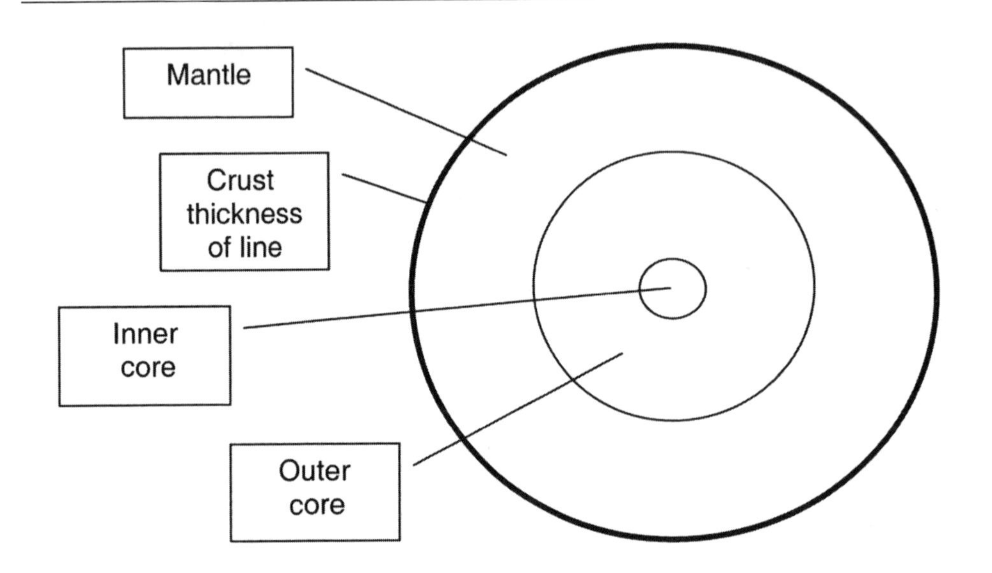

Write three main points from the "Thinking About the Problem" reading:

1. The four main layers of the Earth are ...

2.

3.

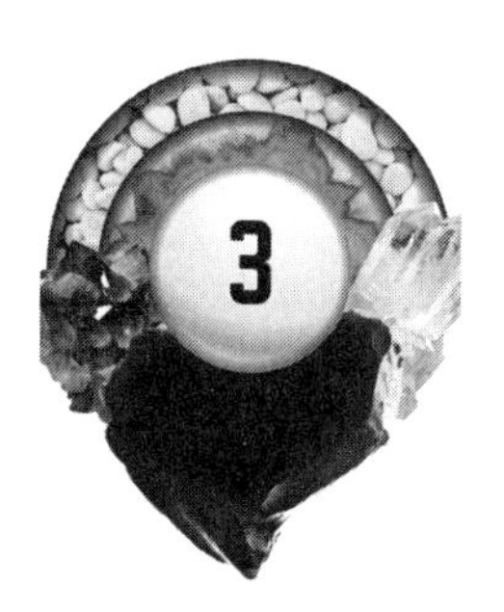

INSERT LABELED SKETCH OF EXPERIMENTAL MATERIALS HERE.

Procedure

1. Sketch a simplified copy of the Hunting Through the Sand Data Table onto a blank sheet of copy paper. It will be used for photos, which will be inserted into your lab in the box above.
2. Place a pinch of sand in the center of your Hunting Through the Sand Data Table. Take a picture of it, and add that picture to the correct spot on your iPad lab report.
3. Use the hand lens to sort the sand grains according to the descriptions on your Hunting Through the Sand Data Table. Take a close-up picture for each spot, once you have separated the grains. Place each image in the correct box on your report (Data Table 3.9).

Analysis

1. Describe, in detail, the various materials that make up your sample of sand.
2. Describe any challenges that you had in sorting your sand grains.
3. (Enrichment) Research the soil profile for your neighborhood (prior to development may be easier to find). Use the internet to find your County/Province Soil Survey. Use the Soil and Water Conservation District or the University Extension Service.

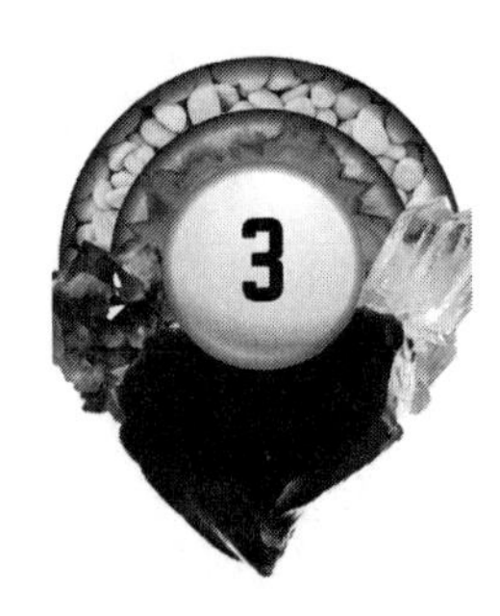

4. (Enrichment) Take a sample of your soil to a gardening store for analysis. Learn what the acidity level is, as well as the percentages of nitrogen, potassium, and phosphorus.

Learning Target

Examine the various materials that make up the sand that is found in a soil sample.

I Learned:

Redo:

Manipulated Variable:

Measured Variable:

Controlled Variable:

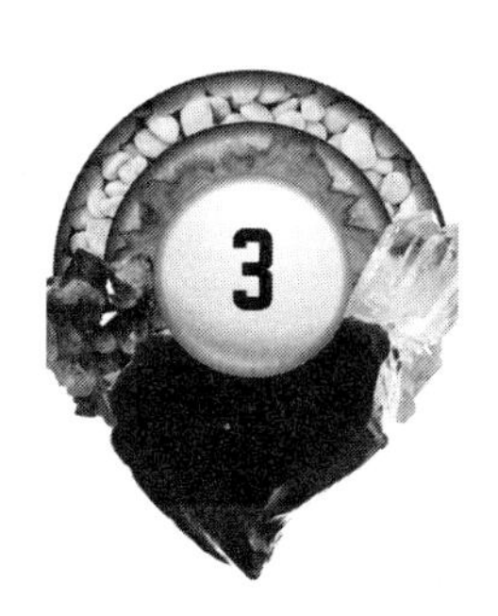

DATA TABLE 3.9.

HUNTING THROUGH THE SAND

Organic Materials (Remains of dead plants)	**Magnetite** (Black or brown, rounded pebbles, magnetic)	**Limestone** (White to light brown, not shiny)
Granite (Different colors, made up of several types of crystals, larger)	**My Sand Box**	**Mica** (Clear to smoky dark, shiny, thin sheets or flakes)
Quartz (Different colors, shiny, looks like glass)	**Slate** (Black to gray, not shiny, tiny angular crystals)	**Feldspar** (White to light brown, jagged/sharp, not shiny)

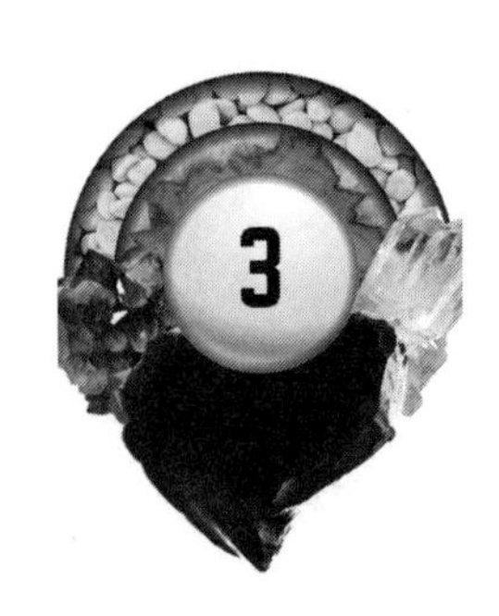

3G. Hunting Through the Sand Lab

NGSS Alignment

MS-ESS2-1. Develop a model to describe the cycling of Earth's materials and the flow of energy that drives this process.

MS-ESS2-2. Construct an explanation based on evidence for how geoscience processes have changed Earth's surface at varying time and spatial scales.

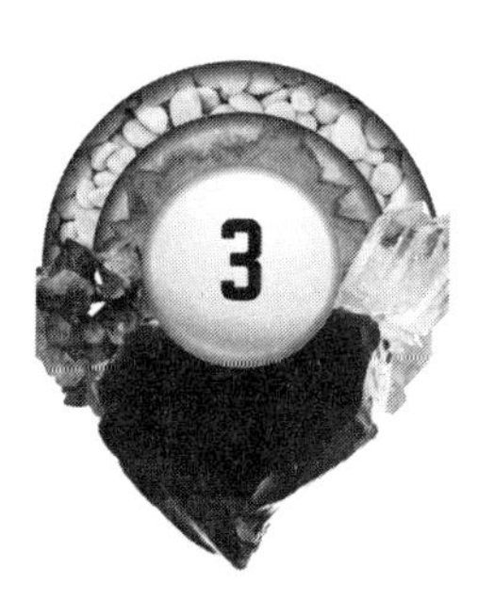

3H.

The Basics of Rocks and Minerals Background Reading

Rocks and minerals make up the Earth around us. But what exactly are rocks and minerals? What is the difference between them? How do they form? Where are they found? To answer these questions, we will begin by defining minerals as, quite simply, the building blocks for making rocks. When you look at a rock and see the different colors and textured surfaces, you are seeing the minerals that make up that specific rock. There are over 3,000 named minerals, but really only about 30 minerals are common.

There are four criteria that must be met in order for something to be called a mineral. First, it must not be formed from the remains of plants or animals (it must be inorganic). Second, it is naturally occurring, not made in a laboratory. Third, it has the same chemical makeup wherever it is found (ex: quartz is always SiO_2). Finally, it has a crystalline structure, which means that it has a specific repeating pattern of atoms.

If all four of the criteria are met, then the substance is a mineral. There are a few tests that geologists rely on to identify which minerals they are looking at: color, shape, a similar form of density called specific gravity, hardness, streak, and luster.

- Color: *Color* is a very common way to try to identify a mineral; however, it should not be used on its own. Because any mineral can be any color, you cannot use color alone to identify a mineral.
- Shape: Some minerals, such as quartz, form in certain *shapes* based on the elements that make them up.
- Density: The mass of a pure mineral, divided by its size (volume) tells you its *density*, unique only to it. Scientists actually use a similar calculation, called specific gravity.
- Hardness: The *hardness* of minerals is based on tests in the Mohs hardness scale, which ranges from 1–10, 1 being the softest and 10 the hardest.
- Streak: The *streak* of a mineral is simply the color of a powder that's left behind when the mineral is scratched along a white, ceramic, unglazed tile.

Even if the color of the mineral itself changes from one specimen to another, the streak color is always the same.

- Luster: *Luster* simply means the way that light reflects off of a mineral, making it dull or shiny.

Rocks are made up of one or more minerals, and solidify into one of three groups. Those groupings of rocks change through the years, however, as Earth processes, such as melting and erosion, affect the rocks. The rock cycle helps to explain how particular rocks are formed from other rocks. All of the rocks on Earth are formed in one of three main ways.

Igneous rocks form when magma or lava cools and hardens into a rock. Magma and lava are rocks that are so hot they move like a liquid. Magma is molten rock that is underground; lava is molten rock that is above the ground.

Sedimentary rocks can form in several different ways. Some sedimentary rocks form from the breakdown material of other rocks. When a rock is weathered, it will begin to erode. The small pieces that break off and erode will collect in oceans and lakes. Over enough time and with enough pressure, these pieces will be compressed and cemented together to form a larger rock, like conglomerate.

Other sedimentary rocks are formed when mineral-rich water evaporates and leaves material behind. An example is halite. Halite is formed when seawater evaporates and leaves behind the salt, which then gets compressed and hardens.

Additionally, some sedimentary rocks are formed from fossils. They can be formed from the hard parts of dead organisms, such as bones and shells, which are cemented together and crystallize into a rock, such as limestone. Or they can be formed when soft parts of living things such as plants die and are buried deep in the Earth, turning them into coal after millions of years of pressure.

Metamorphic rocks are formed when a rock is buried deep in the Earth and is subjected to extreme pressure and heat. The pressure and heat causes the minerals in the rock to undergo chemical changes. This ends up changing the rock into a completely different rock, as when granite changes, or metamorphoses, into gneiss.

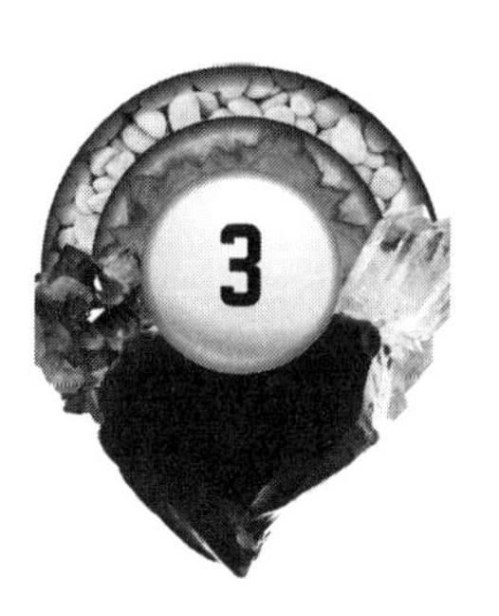

3H. The Basics of Rocks and Minerals Background Reading

NGSS Alignment

MS-ESS1-4. Construct a scientific explanation based on evidence from rock strata for how the geologic time scale is used to organize Earth's 4.6-billion-year-old history.

MS-ESS2-1. Develop a model to describe the cycling of Earth's materials and the flow of energy that drives this process.

MS-ESS2-3. Analyze and interpret data on the distribution of fossils and rocks, continental shapes, and seafloor structures to provide evidence of the past plate motions.

4

HISTORY OF PLANET EARTH

4A. Unearthing History Lab

Problem

Where do life forms appear in a timeline of Earth history?

Prediction

Answer the problem statement.

(*Teacher note:* Have students share several predictions out loud, so misconceptions can be anticipated and explained.)

Thinking About the Problem

How old is Earth? Geologists use information from rocks, rock layers, fossils (*fossus* "dug up" in Latin), and other natural evidence to piece together the history of our planet. Geologists consider time from the formation of the Earth to today, following a geologic timescale that breaks Earth's history into manageable pieces. Geologic time is divided and subdivided into eons, eras, periods, epochs, and ages. They have used this information to put geologic events and fossil organisms (evidence of living things) in their correct sequence on this timeline. The boundaries are set by major events that have been preserved in the rock record.

More recent events can be measured in the soil, as well. For example, Earth systems scientists now believe that an early culture of humans, known as the Clovis people, wandered North America, hunting mammoths and sloths. Their culture came to an end when a mile-wide comet wiped them out. Scientists believe this due to evidence found in a thin layer of black soil, containing iridium from comets, which coats more than 50 sites in North America, especially near the Great Lakes.

Through research, including the use of the geologic time scale, most scientists conclude that Earth is approximately 4.6 billion years old. You will learn more about what evidence scientists use to determine this age in our next lab. Compared to 4.6 billion years, living things have been around for a relatively short time. This lab will help you learn about the geologic timeline for the Earth and more clearly understand the various geological periods and events you will hear described in the media.

Materials

- Earth History on a Rope Scale Model Measurements
- Masking tape
- Rope or twine (5 meters long)
- Ruler
- Scrap paper for labels

Procedure

1. Lay the rope out on the ground in front of you. At the far right end, tape the label "Present Day."
2. Starting from the "Present Day" mark, measure back exactly 4.6 meters. Label this "Formation of the Earth."
3. Measure from the Present Day mark, using Data Table 4.1 (p. 157), and label each eon, era, period, and epoch (with a different color code).
4. Use Data Table 4.2 (p. 158) to label each event in Earth's history.

(*Teacher note:* This is treated as a presentation of "Earth History on a Rope" for Enriched Science, and an "Earth History Timeline" for Regular Science—using a 46 cm line drawn on paper. The four sample analysis questions are for "practice" to prepare for their two presentation questions in front of their peers. As a small group, students work on the rope timeline and you spot-check three particular measurements. If the students are within 2 cm of actual, then they get full credit for the rope timeline. If they are not, then they each lose 2 points for each incorrect measurement (out of 30 points total). Then, for individual accountability, each student has to answer two questions from the list on this document. Each question costs them 2 points per wrong answer. This means that the lowest score any student will receive, should they work in a group on the rope timeline, is 20 out of 30. Data tables include information pertinent to the state of Minnesota.)

Sample Analysis

1. Hypothesize how the geologists divided the time scale into smaller units.
2. Where on the timeline are the two major extinction events?

3. The time from 4.6 billion years ago up until the beginning of the Phanerozoic eon is called Precambrian Time. Find this part of your timeline. How does Precambrian Time compare in length with the rest of the geologic time scale?

4. The Cenozoic era is the most recent era, and it includes the present. How does the Cenozoic era compare in length with the other eras?

Geologic Timescale Presentation Questions

Make questions #1 and #26 available to prepare students, while they get their rope timelines and are waiting to be called on. (*Teacher note:* Answers are given in parentheses.)

1. How many points, or lengths from present day, are marked with events or dates on your rope? (46)

2. Name the eras that are marked on your rope. (Paleozoic, Mesozoic, Cenozoic)

3. Describe what the color code is for the five colors that you used. (Need five colors: eons, eras, periods, epochs, and fossil/events)

4. The time on your geologic time scale from 4.6 billion years ago up until the Phanerozoic eon is called Precambrian time. How does the length of the Precambrian time compare to the rest of the scale? (Precambrian is longer.)

5. The Cambrian period marks the beginning of the complex life forms (like trilobites) in Earth history. How does the length of the Cambrian to present compare with the length of the rest of the geologic time scale? (Cambrian is shorter.)

6. When on the timeline are the two major extinction events? (248 million years ago and 65 million years ago)

7. Hypothesize or explain how you think geologists divided the time scale into smaller units. (Based on life forms found in each unit)

8. Which came first, the Rocky Mountains or the dinosaur extinction? (Rocky Mountains)

9. Which came first, the mammals or the reptiles? (Reptiles)

10. Which came first, the mammals or the flowering plants? (Mammals)

11. Which came first, the amphibians or the reptiles? (Amphibians)

12. Which came first, the flowering plants or the dinosaur extinction? (Flowering plants)
13. Which came first, the Rocky Mountains or the continental ice age being over? (Rockies)
14. Which came first, the green algae or the trilobites? (Green algae)
15. Which came first, the amphibians or the trilobites? (Trilobites)
16. Which era lasted longer, Paleozoic or Mesozoic? (Paleozoic)
17. Give a good Redo Statement for this lab.
18. The Archean eon marks the beginning of the simple life forms (like bacteria) in Earth history. How does the length of the Archean to the beginning of the Cambrian period compare with the length of the Cambrian period to present day? (Archaen to Cambrian is longer.)
19. The Proterozoic eon marks the halfway point for Earth's history. How does the length of the Proterozoic to Cambrian compare with the length of the Cambrian to present day? (Proterozoic to Cambrian is longer.)
20. Which came first, the extinction of dinosaurs or the greatest mass extinction? (greatest mass extinction)
21. Which came first, the mammals or the greatest mass extinction? (Greatest mass extinction)
22. Which came first, the continental ice age or the modern humans? (Modern Humans)
23. Which came first, the Carboniferous period or the reptiles? (Carboniferous)
24. Which came first, the Ordovician period or the first trilobite? (Trilobite)
25. Which came first, the Cenozoic era or the extinction of the dinosaurs? (They're both the same date)
26. Why is the following phrase significant? Pregnant camels often sit down carefully. Perhaps their joints creak… though possibly they're not quick. (First letter of all periods.)
27. Which came first, the Paleozoic era or the Mesozoic era? (Paleozoic)
28. Which period lasted longer, Cambrian or Ordovician? (Cambrian)
29. Which came first, the Carboniferous period or the Silurian period? (Silurian)
30. Which came first, the Milocene epoch or the Eocene epoch? (Eocene)

31. Which lasted longer, the Jurassic period or the Cretaceous period? (Cretaceous)
32. Which Precambrian eon lasted longer, the Priscoan or Archean? (Archean)
33. Which came first, the Triassic period or the Tertiary Paloegene period? (Triassic)
34. Which lasted longer, the Ordovician period or the Silurian period? (Ordovician)
35. During which period are trilobites first found? (Cambrian)
36. During which period were the first mammals found? (Triassic)
37. During which period were the first flowering plants found? (Cretaceous)
38. During which period did the Rocky Mountains begin to rise? (Cretaceous)
39. During which period were the first amphibians found? (Devonian)
40. During which epoch were modern humans first found? (Pleistocene)
41. What event marks the beginning of all of the epochs? (The extinction of the dinosaurs)
42. During which period were the first reptiles found? (Carboniferous)
43. During which eon were the first green algae found? (Precambrian Proterozoic)
44. Which came first, the Ordovician period or the Quaternary period? (Ordovician)
45. Which came first, the Permian period or the Cenozoic era? (Permian)
46. Which eon came first, the Precambrian Priscoan or the Precambrian Archean? (Priscoan)
47. Which came first, the Eocene epoch or the Pliocene epoch? (Eocene)
48. Which Epoch lasted longer, the Oligocene or the Miocene? (Miocene)
49. Which came first, the Carboniferous period or the Permian period? (Carboniferous)
50. Which came first, the continental ice age or the start of the Pleistocene epoch? (Pleistocene)
51. Which epoch lasted longer, the Pliocene or the Miocene? (Miocene)
52. Which era lasted longer, the Mesozoic or the Cenozoic? (Mesozoic)

53. Which event came first, the first green algae or the first bacteria? (Bacteria)
54. Which event came first, the rise of the Rocky Mountains or the first mammal? (Mammal)
55. Which period lasted longer, the Devonian or the Triassic? (Devonian)
56. Which eon lasted longer, the Precambrian Priscoan or the Precambrian Archean? (Archean)
57. Which came first, the Silurian period or the first amphibian? (Silurian)
58. Which period lasted longer, the Tertiary Neogene or the Quaternary? (Tertiary Neogene)
59. Which came first, the Holocene epoch or the end of the continental ice age? (Holocene)
60. Which eon came first, the Precambrian Priscoan or the Precambrian Archean? (Precambrian Priscoan)
61. Which came first, the rocks in Lac Qui Parle, Minnesota, or the rocks in Taylor's Falls, Minnesota? (Lac Qui Parle)
62. Which came first, the inland sea or the glaciers covering Minnesota? (Inland sea)
63. Which came first, the glaciers or the humans? (humans)
64. Which came first, the glaciers or the Minnesota River Valley? (glaciers)
65. Which event came first, the inland sea in Minnesota or the flowering plants? (inland sea)
66. During which eon were the gneiss rocks formed in Lac Qui Parle State Park, Minnesota? (Precambrian Archean)
67. During which period was Minnesota covered by inland seas? (Jurassic)
68. During which period did the Superior Lobe and Des Moines Lobe Glaciers leave deposits in Minnesota? (Quaternary)
69. How would the rope timeline compare in length with one created for Mars? (Both ropes would be the same length)
70. How would the rope timeline compare in length with one for the Moon? (Both ropes would be the same length)

DATA TABLE 4.1.

EARTH HISTORY

GEOLOGISTS' DIVISION OF EARTH HISTORY	HOW MANY MILLIONS OF YEARS AGO IT BEGAN	MEASUREMENT ON ROPE (0.1 CM = 1 MILLION YEARS)
Chronometric Eons		
Precambrian Priscoan	4600	460.0 cm
Precambrian Archean	3800	
Precambrian Proterozoic	2500	
Phanerozoic	544	
Eras		
Paleozoic	544	54.4 cm
Mesozoic	248	
Cenozoic	65	
Periods		
Cambrian	544	
Ordovician	490	49.0 cm
Silurian	443	
Devonian	417	
Carboniferous	354	
Permian	290	
Triassic	248	
Jurassic	206	
Cretaceous	144	
Tertiary Paleogene	65	
Tertiary Neogene	24	
Quaternary	2	
Epochs		
Paleocene	65	6.5 cm
Eocene	55	
Oligocene	34	
Miocene	24	
Pliocene	5	
Pleistocene	2	
Holocene	0.01	

DATA TABLE 4.2.

EVENTS IN EARTH'S HISTORY

EVENTS	TIME (MILLIONS OF YEARS AGO)
Continental ice age is over in United States	0.001
Glacial river Warren carves out the Minnesota River Valley	0.0012
Superior Lobe and Des Moines Lobe Glaciers leave deposits in Minnesota	0.002
Modern humans	0.5
Early humans	2
Extinction of dinosaurs	65
Rocky Mountains begin to rise	80
Flowering plants	130
Twin Cities are covered by seas	150
First mammal	210
Greatest mass extinction	248
First reptiles	315
First amphibians	367
Inland Sea covers Minnesota	480
Minnesota is positioned over the equator	300
First trilobite	554
First green algae	1000
Basalt rocks formed in Taylor's Falls, Minnesota	1100
Gneiss rocks formed in Lac Qui Parle State Park, Minnesota	3600
First bacteria	3800

4A. Unearthing History Lab

NGSS Alignment

MS-ESS1-4. Construct a scientific explanation based on evidence from rock strata for how the geologic time scale is used to organize Earth's 4.6-billion-year-old history.

MS-ESS2-3. Analyze and interpret data on the distribution of fossils and rocks, continental shapes, and seafloor structures to provide evidence of the past plate motions.

4B.

Drilling Through the Ages Lab

Problem

How can we use drilling for wells as a way to understand geologic history?

Prediction

What methods can be used to determine the ages of rock layers?

(*Teacher note:* Have students share several predictions out loud, so misconceptions can be anticipated and explained.)

Thinking About the Problem

Why are geologists interested in drilling? Geologists work together with engineers when drilling for groundwater wells. Drilling allows geologists to examine where different layers of rock begin and end. In the search for water, geologists frequently look for a layer of sandstone perched above a layer of impermeable shale.

Geologists also have an interest in drilling because rock layers provide a record of events that have occurred on Earth. They can contain the remains and imprints of the different plants and animals that have lived on Earth. There are many deep wells (water, oil, and so on) available for geologists to examine.

Scientists estimate that Earth is approximately 4.6 billion years old. There are many pieces of supporting evidence for this. One piece of supporting evidence is the thickness of the rock layers on Earth. Scientists can perform experiments to determine how long it takes to create one meter of a particular rock type. They then multiply this time by the actual thickness of those particular rock layers on Earth. This allows scientists to roughly estimate the age of the Earth. Most geologists believe that it would have taken approximately 4.6 billion years to generate all the layers of rock found on Earth. This study of rock layer depths has been backed up by much more accurate evidence from radioactive minerals and index fossils in the rocks.

Earth scientists study the evidence associated with when the continents began to solidify. Newly discovered Greenland outcrops (an ancient piece of the sea floor, which was raised up by crustal movement) are among the oldest measured, at 3.8

billion years, while most of the continents are much younger, at 2.5 billion years old.

By understanding some simple rules about rock layer formation, we can use the layers and the associated rock types to measure the amount of time that has passed. One important thing to remember is that rock layers form horizontally. A second important factor is that the older rocks will normally be found farther beneath the surface, while younger rocks will normally be closer to the top. This allows scientists to use the positions underground to determine the "age based on position."

Scientists can use index fossils to determine the "relative age" of layers. Index fossils are the remains of a single species that are so widespread and well known (age-wise), that its fossils enable geologists to correlate environments and time. They can also measure the radioactive minerals found in a rock layer to determine the "absolute age" of the layer.

Write three main points from the "Thinking About the Problem" reading:

1.

2.

3.

Procedure

1. At each drilling site on Figure 4.1 (p. 163), place a small horizontal line at the depths described in Data Table 4.3 (p. 162). Write the name of the rock on that line. The first line for Water Well C, sandstone, has been done for you.
2. Draw a line across the page to connect the areas on all three wells where the rock layers are the same.
3. Use the notes from Data Table 4.3 to determine the age of each rock layer. Write the age in parentheses to the right of the rock layer name.
4. Complete Data Table 4.4 (p. 164) in order from youngest (1) to oldest (11).

DATA TABLE 4.3.

INFORMATION FROM WATER WELLS A, B, AND C

WATER WELL A		
DEPTH (M)	**ROCK**	**GEOLOGIST NOTES**
12	Shale	
16	Conglomerate	
25	Sandstone	135 million years old (Index fossils found)
30	Impermeable Shale	No Date Available
45	Breccia	
53	Sandstone	
58	Shale	
WATER WELL B		
DEPTH (M)	**ROCK**	**GEOLOGIST NOTES**
15	Shale	21 million years old (Index fossils found)
16	Conglomerate	
23	Sandstone	
26	Impermeable Shale	
45	Breccia	
51	Sandstone	280 million years old (Radioactive dating)
60	Shale	310 million years old (Index fossils found)
70	Schist	385 million years old (Radioactive dating)
76	Marble	
85	Basalt	
WATER WELL C		
DEPTH (M)	**ROCK**	**GEOLOGIST NOTES**
5	Sandstone	0.5 million years old (Radioactive dating)
18	Shale	
21	Conglomerate	51 million years old (Index fossils found)
25	Sandstone	
34	Impermeable Shale	
47	Breccia	230 million years old (Index fossils found)
55	Sandstone	
63	Shale	
70	Schist	
75	Marble	405 million years old (Radioactive dating)
81	Basalt	460 million years old (Radioactive dating)

FIGURE 4.1.

DRILLING THROUGH THE AGES DIAGRAM

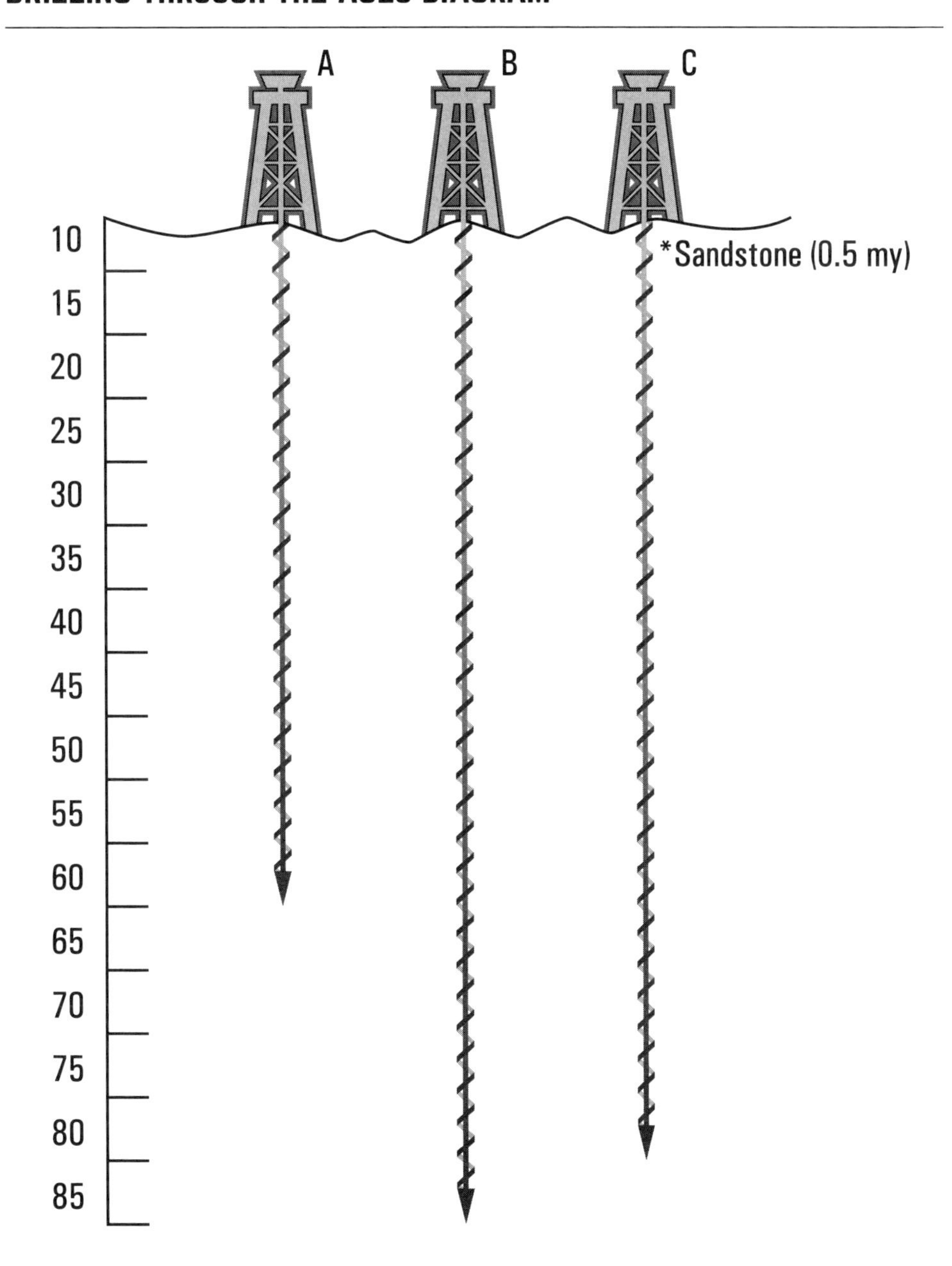

DATA TABLE 4.4.

AGES OF EACH ROCK LAYER

NUMBER OF ROCK LAYER	ERA	PERIOD	AGE OF ROCK LAYER	METHOD USED TO DETERMINE AGE	TYPE OF ROCK
1	Cenozoic	Quaternary	0.5 million years	Radioactive	Sandstone
2					
3					
4					
5					
6					
7					
8					
9					
10					
11					

DATA TABLE 4.5.

GEOLOGIC TIME TABLE

MILLIONS OF YEARS AGO	ERA	PERIOD
0–2	Cenozoic	Quaternary
2–24	Cenozoic	Tertiary Neogene
24–65	Cenozoic	Tertiary Paleogene
65–141	Mesozoic	Cretaceous
141–195	Mesozoic	Jurassic
195–230	Mesozoic	Triassic
230–280	Paleozoic	Permian
280–310	Paleozoic	Pennsylvanian
310–345	Paleozoic	Mississippian
345–395	Paleozoic	Devonian
395–435	Paleozoic	Silurian
435–500	Paleozoic	Ordovician
500–570	Paleozoic	Cambrian

Analysis

1. Explain one other way to find out the age of the rock in layer #5.
2. Explain whether or not all of the similar rock types are found at the same depth.
3. Describe the difference between the relative age of a rock layer and its absolute age.
4. (Enrichment) What type of evidence would be found in the rock layers if there had been volcanic eruptions in the past?
5. (Enrichment) Construct a scientific explanation about how you would use evidence found in rock layers, in order to prove that volcanic eruptions had happened in the past.
6. (Enrichment) Contact a local water, oil, or gas drilling company to discover and learn from their methods of collecting evidence on the rock layers underground in our community. Create an iMovie trailer to share with the class to teach them what you've learned.

Learning Target

Use the rock layers and the associated rock types to measure the amount of time that has passed.

I Learned:

Redo:

Manipulated Variable:

Measured Variable:

Controlled Variable:

4B. Drilling Through the Ages Lab

NGSS Alignment

MS-ESS1-4. Construct a scientific explanation based on evidence from rock strata for how the geologic time scale is used to organize Earth's 4.6-billion-year-old history.

MS-ESS2-1. Develop a model to describe the cycling of Earth's materials and the flow of energy that drives this process.

MS-ESS2-2. Construct an explanation based on evidence for how geoscience processes have changed Earth's surface at varying time and spatial scales.

MS-ESS2-3. Analyze and interpret data on the distribution of fossils and rocks, continental shapes, and seafloor structures to provide evidence of the past plate motions.

4-ESS1-1. Identify evidence from patterns in rock formations and fossils in rock layers for changes in a landscape over time to support an explanation for changes in a landscape over time.

4C. Decaying Candy Lab

Problem

How many half-lives will it take for a sample of candy to decay?

Prediction

Give a working definition of "half-life."

(*Teacher note:* Have students share several predictions out loud, so misconceptions can be anticipated and explained.)

Thinking About the Problem

When are rocks born? How do we know what their birthdays are? For igneous rocks, that birthday is when they first harden from magma or lava to become rock. All of the elements within an igneous rock help us to identify it. Most elements within the rock are stable and remain the same through the years. Some, however, are unstable. Over time, these elements decay, or break down, changing into new elements by releasing energy and subatomic particles. This process is called radioactive decay. Radioactive elements, such as uranium and radon, occur naturally in igneous rocks.

Unstable elements are said to be radioactive. During radioactive decay, the atoms of one element break down to form atoms of another element. As a radioactive element within the igneous rock decays, it changes into another element. So the composition of the rock changes slowly over time. The amount of the radioactive element decreases, while the amount of the newly formed element increases.

The particular rate of decay for each radioactive element never changes, and is referred to as the half-life. The half-life measures how long it takes for any quantity of radioactive elements within the rock to decay by half. Geologists use the rate at which these elements decay to calculate the rock's age. They can use radioactive dating to determine what is called the absolute age, or the birthday, of rocks.

As all plants and animals grow and travel through their lives, carbon atoms are added to their tissues. There is a radioactive form of carbon called carbon-14.

All living things contain carbon atoms, including some carbon-14. It has a shorter half-life (5,730 years) than the elements found in igneous rocks, and can be used to determine the age of some living things. After an organism dies, no more carbon is added to the tissues. But since the carbon-14 in the organism's body is radioactive, it decays. It breaks down into a stable nitrogen-14 atom. To determine the age of a once-living thing, scientists measure the amount of carbon-14 that is left in the living thing's remains. From this amount, they can determine the absolute age, or years that have passed since its birthday. Carbon-14 has been used to determine the age of frozen mammoths and prehistoric humans.

Write three main points from the "Thinking About the Problem" reading:

1. Geologists use radioactive dating to ...

2.

3.

Procedure

1. Place 50 "atoms" of candy (M&Ms) in the cup, and gently shake for 10 seconds, representing its half-life.
2. Gently pour out candy. Count the number of pieces with the M&M side up. These atoms have "decayed." Record amount in Data Table 4.6.
3. Return only the pieces with the print-side down to the cup. You may consume the "decayed' (print-side up) atoms.
4. Continue gentle 10-second shaking, counting, and consuming until all the atoms have decayed. Draw a sketch of you materials in box on p. 169.
5. Combine all of the class data, and graph the whole-class average data (Data Table 4.7).
6. In Figure 4.2 (p. 171), label time (seconds) on the *x*-axis. Label the number of undecayed atoms on the *y*-axis. Give your line graph a descriptive title.

INSERT LABELED SKETCH OF EXPERIMENTAL MATERIALS HERE.

DATA TABLE 4.6.

SMALL-GROUP DATA

HALF-LIFE (SECONDS)	# OF UNDECAYED ATOMS (RUNNING TOTAL)	# OF DECAYED ATOMS (RUNNING TOTAL)
0	50	0
10		
20		
30		
40		
50		
60		
70		
80		

DATA TABLE 4.7.

WHOLE CLASS-DATA ON UNDECAYED ATOMS

HALF-LIFE (SEC)	0	10	20	30	40	50	60	70	80
Group 1	50								
Group 2	50								
Group 3	50								
Group 4	50								
Group 5	50								
Group 6	50								
Group 7	50								
Average	50								

FIGURE 4.2. ____________________

(*Teacher note:* Student provides descriptive graph title.)

Three Graphing Hints for Students

1. Read Procedure #6 again.
2. Use the full graph—it should be a big picture of important data. Count by an appropriate number.
3. Use the line on your graph to determine what the half-life of the candy sample is, in seconds.

Analysis

1. Give a working definition of half-life.
2. In the experiment, what was the half-life of the candy sample?
3. At the end of two half-lives what fraction (or percent) of the atoms had not decayed?
4. (Enrichment) How good is our assumption that half of the candy atoms will decay in each half-life? Explain.
5. (Enrichment) Is there any way to predict when a particular atom of candy will decay? (If you could follow the fate of one M&M, is there any way to predict exactly when it will "decay?") Explain.
6. Describe the shape of the curve drawn in on your graph of the class data.
7. Why did we combine to get the whole-class data? How does this relate to radioactive dating?
8. If you started with a sample of 600 atoms of candy, how many would remain undecayed after three half-lives?
9. (Enrichment) If 175 undecayed nuclei remain from a sample of 2,800 nuclei, how many half-lives have passed?
10. (Enrichment) The element Strontium-90 has a half-life of 28.8 years. If you start with a 10 g sample of Strontium-90, how much will remain after 115.2 years? Show your math.

Learning Target

Use the radioactive half-life of elements to determine a sample's absolute age.

I Learned:

Redo:

Manipulated Variable:

Measured Variable:

Controlled Variable:

4C. Decaying Candy Lab

NGSS Alignment

MS-ESS1-4. Construct a scientific explanation based on evidence from rock strata for how the geologic time scale is used to organize Earth's 4.6-billion-year-old history.

MS-ESS2-3. Analyze and interpret data on the distribution of fossils and rocks, continental shapes, and seafloor structures to provide evidence of the past plate motions.

MS-PS1-1. Develop models to describe the atomic composition of simple molecules and extended structures.

4D.

Superposition Diagram Challenge

Directions

With the Law of Superposition, your goal is to determine, and find evidence for, the sequencing of different events in the rock layers of the Earth. Now you will use those sequencing skills to determine the relative ages of various rock layers in a diagram.

Determine the relative age (oldest to youngest) of the rock layers and surfaces in Figure 4.3, labeled A through J, shown in a cross-section diagram of Earth.

Each rock layer or surface is associated with an event that caused it to develop. Some surfaces may have developed due to erosion and weathering. Some rock layers may have developed due to volcanic activity underground. Some may have developed due to deposition and cementation as a sedimentary rock layer. There are many reasons why rock layers develop.

Give clear evidence for why you believe each particular rock layer or surface is older or younger than surrounding layers or surfaces. You will need to be confident with your evidence, so you can defend your choice to others.

Complete Data Table 4.8 (p. 176) to show your sequence and evidence.

FIGURE 4.3.

SUPERPOSITION CHALLENGE

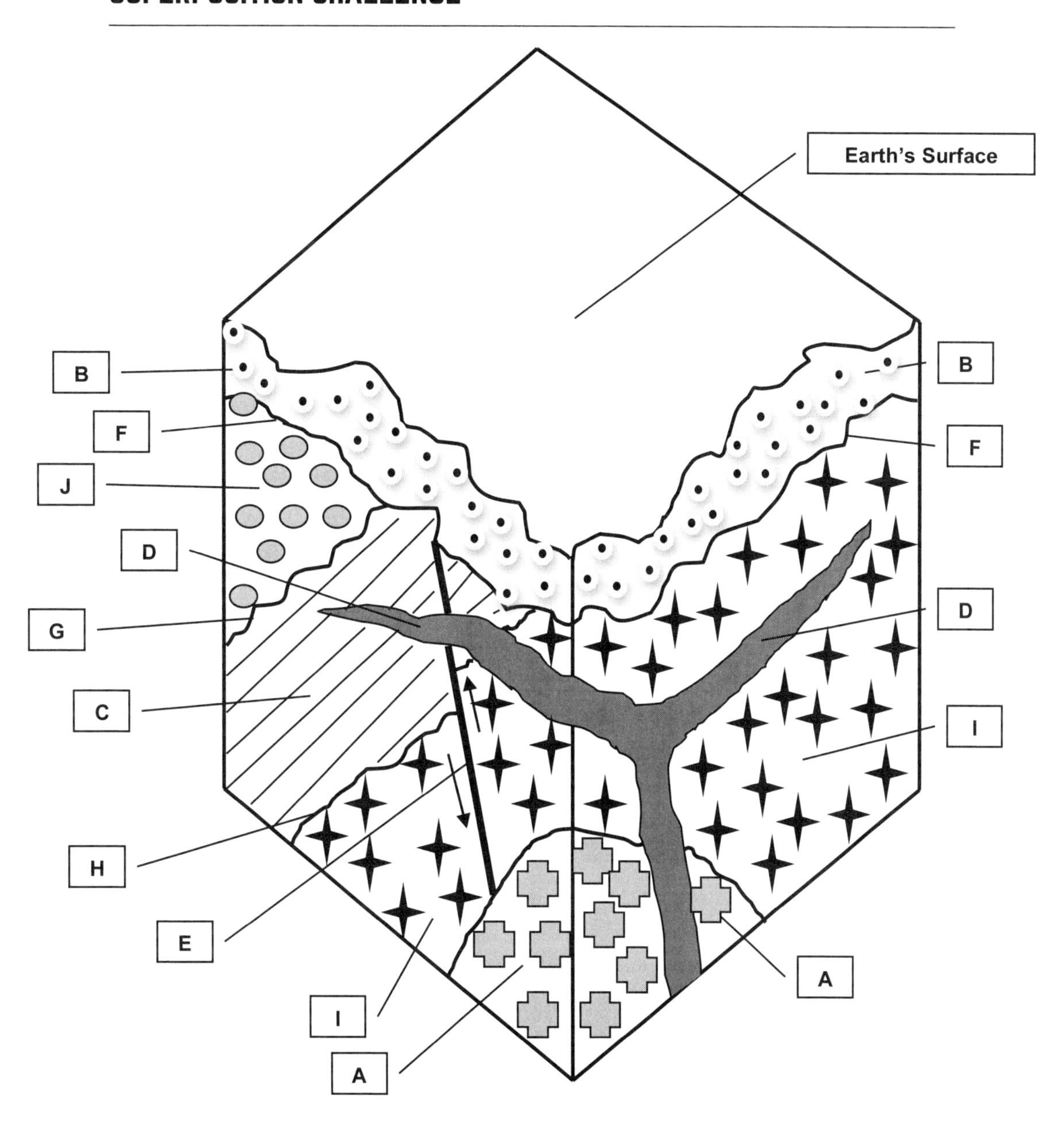

DATA TABLE 4.8.

DATA ON RELATIVE AGE OF ROCK UNITS AND SURFACES A THROUGH J

YOUR HYPOTHESIS (RANKING FROM OLDEST TO YOUNGEST)	BRIEF EVIDENCE	ALTERNATE HYPOTHESIS #1 (FROM ANOTHER GROUP)	ALTERNATE HYPOTHESIS #2 (FROM ANOTHER GROUP)	ALTERNATE HYPOTHESIS #3 (FROM ANOTHER GROUP)	QUESTIONS THAT REMAIN OR OTHER EVIDENCE NEEDED

4D. Superposition Diagram Challenge

NGSS Alignment

MS-ESS1-4. Construct a scientific explanation based on evidence from rock strata for how the geologic time scale is used to organize Earth's 4.6-billion-year-old history.

MS-ESS2-1. Develop a model to describe the cycling of Earth's materials and the flow of energy that drives this process.

MS-ESS2-2. Construct an explanation based on evidence for how geoscience processes have changed Earth's surface at varying time and spatial scales.

MS-ESS2-3. Analyze and interpret data on the distribution of fossils and rocks, continental shapes, and seafloor structures to provide evidence of the past plate motions.

4-ESS1-1. Identify evidence from patterns in rock formations and fossils in rock layers for changes in a landscape over time to support an explanation for changes in a landscape over time.

4E.

Mapping the Glaciers Lab

Problem

How were Minnesotan landscapes affected by glaciers?

Prediction

Give a complete sentence answer to the problem statement above.

(*Teacher note:* Have students share several predictions out loud, so misconceptions can be anticipated and explained.)

Thinking About the Problem

Glaciers are composed of fallen snow that accumulates over time and compresses into thickened ice masses. As the snow piles up, ice begins to form on the bottom of the snow mass. As the mass becomes greater, the ice on the bottom of the snow begins to melt and the pile begins to slide. At this point the snow mass is known as a glacier.

There have been several major ice advances during the last two million years. During these ice advances, glaciers covered large parts of Minnesota. The last of these glacial advances ended about 11,000 years ago.

These glaciers were enormous sheets of ice over a mile thick and covering thousands of square miles. They exerted massive forces due to their tremendous weight. As the glaciers melt, they release a torrent of meltwater, which carved many of the surface landforms visible in Minnesota today. Many of our 12,000 lakes, the Minnesota River Valley, The Great Lakes, and the deep potholes of Taylor's Falls all owe their existence to glacial processes.

In nature, glaciers can move quickly or slowly downslope, depending on the angle of slope, changes in atmospheric temperature and glacial load. As a glacier moves over the land, it "plucks" rock and soil debris from the surface. These plucked rocks become embedded in the ice and act as cutting tools, which in turn smooth and polish the rock surfaces beneath the moving glacier. From time to time, the glacier will drop rocks and sediments to produce glacial landforms, such as boundary piles, called moraines.

Glacial deposits are generally grouped into two classes: till and outwash. Till occurs when different-size sediments (such as sand, clay, and boulders) are deposited from a glacier. Outwash is a glacial deposit of sediment left from the melting ice of a glacier. The melted water from the glacier caries sediments and creates channels in the same manner as a river or stream.

Write three main points from the "Thinking About the Problem" reading:

1. Glaciers are composed …

2.

3.

Procedure

1. Use 3-D satellite image maps for the Upper Midwest, obtained from a local United States Geological Survey (USGS) Office. These maps can also be downloaded and printed from the *USGS.gov* website. Research glacier terms on the internet for any information gaps that remain. (*Teacher note:* The author recommends laminating the maps, so the maps can be marked on and reused.)

2. As a group, work with the satellite image 3-D map. Begin by outlining and labeling the states in one color.

3. Using a hand lens if necessary, find two different land formations that you believe might have been formed by glaciers scraping surfaces, or by glaciers piling up debris. Trace the outlines and label, in a new color, of where you believe the glaciers likely generated these scrape marks or piles. Write your evidence for these glacial land formations in Data Table 4.9 (p. 180).

4. As a group, find two different land formations that you think might have been formed by the flow of water as glaciers melted. Circle and label the two landforms that you believe resulted from these glacial events. Be able to give evidence, as well as detailed descriptions of where those landforms are found. List your evidence in Data Table 4.9 (p. 180).

5. As a group, find two different places on the map where you believe there is evidence showing which direction the glaciers were moving. Determine the direction in which you believe the glaciers traveled, and show it by arrow on the map. Be able to give evidence for your selection. List your evidence in Data Table 4.9.

DATA TABLE 4.9.

EVIDENCE FROM THE SATELLITE IMAGE 3-D MAP

Evidence for Procedure #3	
Evidence for Procedure #4	
Evidence for Procedure #5	

6. What landforms are created by glacial activity? Read the description of each of the formations below. Then find one spot on the map where this type of formation seems to be present. It is okay to repeat spots from earlier. Label them as described below.

   a. Lateral Moraine: Lateral moraines form on the side of the glacier. A moraine is a large pile of glacial debris left behind by a glacier. These formations mark the edges of where the glacier once was, similar to

a riverbank. Circle the spot where you think a lateral moraine has formed and label it "Lateral Moraine."

b. Terminal (End) Moraine: The terminal moraine marks the farthest advance of the glacier. This type of moraine forms along the front of the glacier, similar to how a bulldozer pushes material in front of it. Circle the spot where you think a terminal moraine has formed and label it "Terminal Moraine."

c. Kettle Lakes: As glaciers melt and break up, they often leave behind large blocks of ice partially buried in the ground. As the ice blocks melt, depressions remain in the landscape. The depressions fill with snowmelt and rainwater to produce kettle lakes. In the satellite view, this will look like a region with many small, shallow lakes. Circle the area where you think kettle lakes are located and label it "Kettle Lakes."

7. Now that your group has labeled all of the formations, take a clear picture of your 3-D map to show where you circled evidence. Label the image, and place it in the box below. Once your group has their map approved, use a damp paper towel to clean your map.

LABELED IMAGE OF COMPLETED SATELLITE 3-D MAP

9. Compare and contrast snow banks (in the school parking lot in winter) with glaciers. Complete the thinking map in Figure 4.4.

(*Teacher note:* A lesson called "Comparing and Contrasting Thinking Maps" in Appendix D on page 317, helps students learn how to complete the thinking map.)

FIGURE 4.4.

THINKING MAP: COMPARING AND CONTRASTING GLACIERS WITH SNOW BANKS

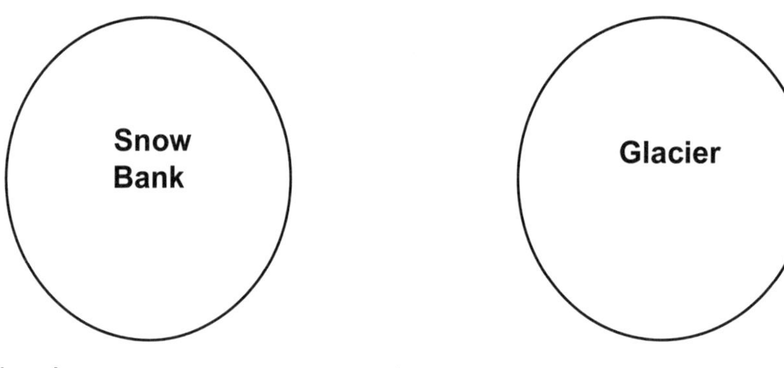

Analysis

1. What is one important observation about how North American landforms were affected by glaciers?
2. Summarize Figure 4.6 by giving one similarity and one difference between snow banks and glaciers:
   - Similarity:
   - Difference:
3. How does weather play a role in changing glaciers and snow piles?
4. What could happen to the ground underneath glaciers and snow piles?
5. What do you know about glaciers in Minnesota now that you have spent time observing evidence for them closely?
6. (Enrichment) Use what you know about glaciers to describe in detail exactly what will happen to the snow piles in our school parking lot, over time as they melt.

7. (Enrichment) Some landforms are too small to be visible on a large 3-D map. Conduct some internet research to determine how drumlins, eskers, and till are formed.

Learning Target

Use maps, satellite images, and other data sets to describe patterns and make predictions about local and global systems involving glaciers.

I Learned:

Redo:

Manipulated Variable:

Measured Variable:

Controlled Variable:

4E. Mapping the Glaciers Lab

NGSS Alignment

MS-ESS1-4. Construct a scientific explanation based on evidence from rock strata for how the geologic time scale is used to organize Earth's 4.6-billion-year-old history.

MS-ESS2-2. Construct an explanation based on evidence for how geoscience processes have changed Earth's surface at varying time and spatial scales.

4F.

Geoarchaeology Background Reading

How do we know what the Earth was like millions of years ago? Geoarchaeology is a science that uses evidence derived from the application of Earth sciences to archaeological problems. The physical landscape of the Earth is the setting in which human activities have taken place. Geoarchaeology has contributed to our understanding of the human past. It has shown us evidence for environmental change and the evolution of Earth's physical landscape. It has also shown us the processes that affect the formation, preservation, and destruction of archaeological sites.

Geoarchaeology derives its evidence from the shape of the landscape, deposits within the landscape, and soils developed across it. Major changes in the physical landscape—continents, mountain chains and oceans—mostly take place on geological timescales of millions of years. Smaller-scale changes in, for example, the position of rivers and coastlines, however, occur on the shorter archaeological timescale. Such changes in the shape of the physical landscape are studied by the science of geomorphology (literally the study of "earth shapes") and are of importance to archaeology at the landscape scale in several ways.

How do we know what happened long ago? First, by observing modern processes, the environments in which they take place, the landforms which they produce, and the fossils present, we can infer the processes that operated in the past. The shape and structure of features enables us to reconstruct past environments using a key geology idea, the principal of uniformitarianism, which states that, "the present is the key to the past."

Second, understanding of the physical landscape is a requirement for reconstructing the prehistoric geography of a region. The structure of the landscape affects the types of activities that human groups can carry out with a reasonable expectation of success—dictating the types of soils present (potential for agriculture), water supplies, the viable communication routes (rivers and mountain passes), and the defensibility of sites.

Third, the survival of archaeological evidence is affected by modern-day processes, which can destroy sites or cover them so they are no longer visible at the surface. Scientists need to have some knowledge of how the landscape has changed from the historic period under study through current day.

Geoarchaeologists describe the structure of landscapes in terms of three main things: bedrock, drift geology, and landforms.

Bedrock is the foundation rock. The composition of rocks in any area will affect their resistance to weathering and erosion. Harder, more compact rocks (granite and basalt) tend to resist weathering and erosion better than loosely consolidated ones (sandstone and chalk). The harder rocks will tend to form high-standing features. The composition of the bedrock will also affect the nature of the soils developed on its surface.

Drift geology involves any unconsolidated sediment that lies on top of the bedrock. These sediments are usually of geologically recent origin—dating from the Quaternary epoch. They comprise sediments such as boulder clays or tills (deposited by former ice sheets), terrace gravels (deposited by ancient rivers), and marls or clays (deposited in former lake basins).

Landforms are the shape of the physical landscape. Landforms may be described by their mode of origin as either erosional or depositional. Erosional landforms are those features that are cut into the surface of the bedrock and/or drift geology and include features such as the U-shaped troughs of glaciated valleys and cliffs cut by the action of the sea. Depositional landforms are those that are built up and are composed of deposits. They include river terraces, drumlins, kames, eskers, and end moraines.

In all cases it is a combination of the shape of a feature, its composition, and its internal structure that enables description and interpretation of a landform's origin. It is important to distinguish between those processes that result in the breakdown, disintegration, and alteration of rocks and sediments (collectively termed "weathering"), as well as those that result in the transportation and movement by gravity, wind, and water of material that has previously been weathered (collectively termed "erosion"). The study of the physical landscape and the processes operating within it, geoarchaeology, is essential to all of these processes.

4F. Geoarchaeology Background Reading

NGSS Alignment

MS-ESS1-4. Construct a scientific explanation based on evidence from rock strata for how the geologic time scale is used to organize Earth's 4.6-billion-year-old history.

MS-ESS2-1. Develop a model to describe the cycling of Earth's materials and the flow of energy that drives this process.

MS-ESS2-2. Construct an explanation based on evidence for how geoscience processes have changed Earth's surface at varying time and spatial scales.

5
EARTH'S INTERIOR SYSTEMS

5A.

Shaking Things up Lab

Problem

Why do earthquakes happen in certain locations around the world?

Prediction

Describe, in one sentence, where you think most earthquakes in the world will occur over the next month.

(*Teacher note:* Have students share several predictions out loud, so misconceptions can be anticipated and explained.)

Thinking About the Problem

Can you predict earthquakes? Scientists have never predicted a major earthquake. They do not know how, and they do not expect to know how any time soon. However, probabilities, based on scientific data, can be calculated for potential future earthquakes. Scientists are becoming better and better at predicting the likelihood of potential earthquakes.

Plates are the slabs of the Earth's crust that make up the lithosphere (*lithos,* "stone" in Greek). Geologists developed the plate tectonic theory as a model of movement of Earth's crust on the surface. Earth's crust is composed of the continental crust (30 to 100 km thick) and the oceanic crust (about 10 km thick). Faults (*fallere,* "to fail" in Latin) are fractures in the Earth's crust caused by the stresses of plate movements.

The word *earthquake* is used to describe any seismic (*seismos,* "earthquake" in Greek) event—whether a natural phenomenon or something caused by humans—that generates seismic waves. Seismic waves are caused mostly by the rupture of geological faults, but also by volcanic activity, landslides, mine blasts, fracking, and nuclear experiments. An earthquake is usually the result of a sudden release of energy in Earth's crust, due to slippage along geologic faults, causing a vibration of the Earth.

There are many misconceptions about earthquakes. Some believe that animals can predict earthquakes. Unresearched evidence does exist of animals, such as fish, birds, reptiles, and insects, exhibiting strange behavior anywhere from weeks to seconds before an earthquake occurs. However, consistent and reliable behavior prior to seismic events has never been shown.

Earthquakes may occur near a volcanic eruption, but they are the result of the active forces connected with the eruption, and not the cause of volcanic activity. As well, contrary to some beliefs, earthquakes are equally as likely to occur at any time of the day, month, or year.

Geologists use the Moment Magnitude Scale (MMS) to assign magnitude to earthquakes by the height of the largest seismic wave that each earthquake creates. Each unit of additional magnitude refers to a tenfold increase in the level of ground shaking and an even more dramatic increase in energy. For example, a magnitude 7.0 earthquake has over 30 times more energy than a magnitude 6.0. The MMS was developed in the 1970s to replace the Richter scale. The MMS is used by the United States Geological Survey to rate the scale of earthquakes.

Write three main points from the "Thinking About the Problem" reading:

1.

2.

3.

Materials

- Colored pencils
- Internet access to the United States Geological Survey's Earthquake Page
- Wall-size map of the World (such as in Figure 5.1)

FIGURE 5.1.

MAP OF THE WORLD

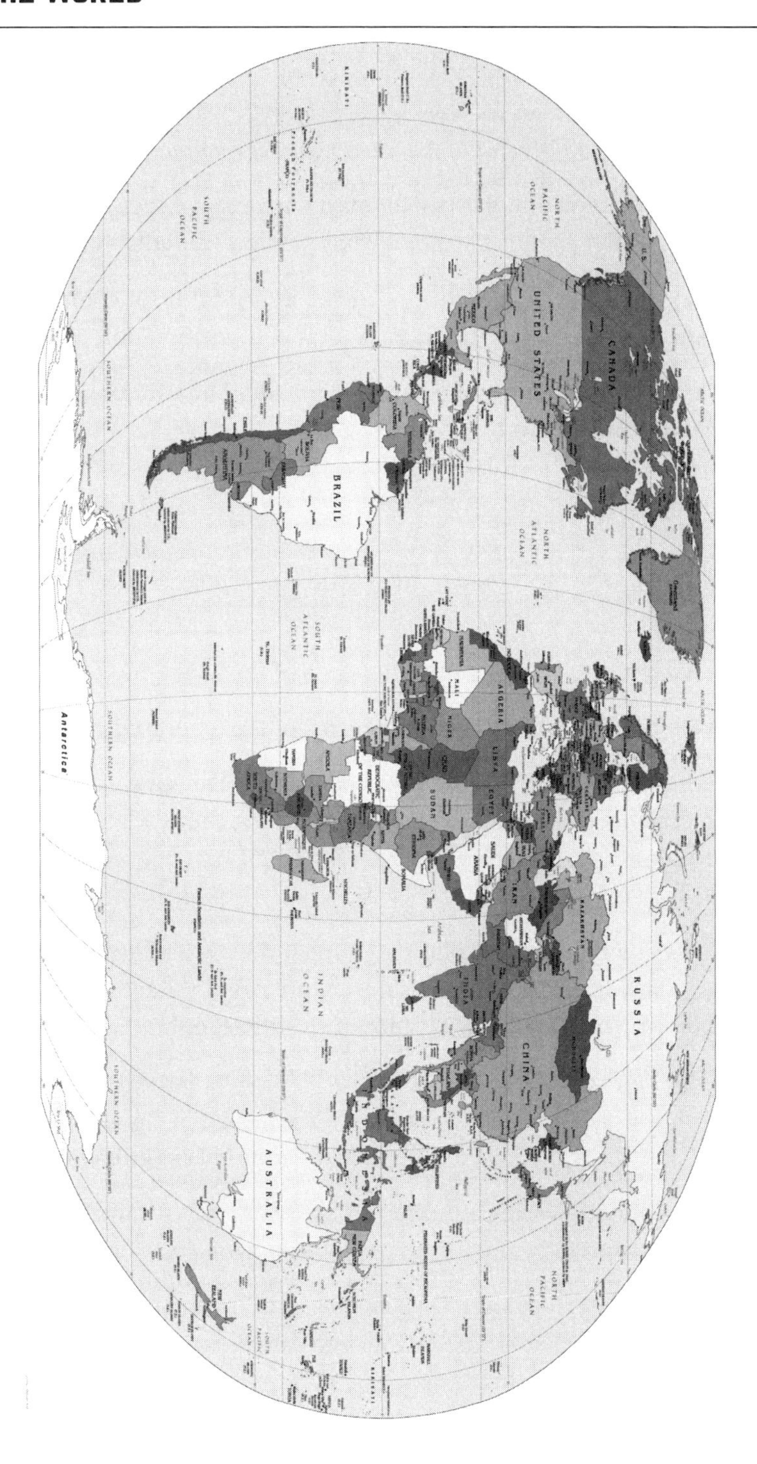

Procedure

1. Post a map of the world on a wall in the classroom.
2. Students should have a smaller world map on their lab report in *Notability*.
3. Visit the United States Geological Survey's "Current Worldwide Earthquake List" at *http://earthquake.usgs.gov/regional/neic*.
4. Place a colored dot on the wall map where any earthquakes have occurred that day. Repeat this pattern weekly (assign students this job).
5. Students use the highlighting feature in *Notability*, or colored pencils to indicate the same information in their composition notebooks.
6. Use different colors to indicate magnitudes of earthquakes (strong, medium, and weak).

Analysis

1. Describe how your map of the world has changed over the course of four months.
2. What patterns do you notice in terms of the locations of earthquakes?
3. What patterns do you notice in terms of the magnitudes of earthquakes?
4. (Enrichment) Build the tallest building you can, given only 20 index cards and one meter of masking tape. Test it for stability. Compete with other lab groups for height. The winning lab group is the one with the tallest tower that can also withstand the "shaking desk" quake.
5. (Enrichment) Visit the following link and view a plate tectonics animation from Howard Hughes Medical Institute's BioInteractive site (*http://goo.gl/LgIOsn*). Then write one "I learned …" statement.
6. (Enrichment) Students used different colors to indicate magnitudes of earthquakes for the main lesson. Follow a similar procedure to trace subduction zones, by noticing depths of earthquakes (in 100 km increments).
7. (Enrichment) Research and report on a fault zone in the United States, sharing especially any future predictions for activity levels.
8. (Enrichment) Develop a *Keynote* presentation on building materials and methods that are designed to sustain earthquakes without damage.

Learning Target

Monitor the locations of earthquakes as they occur around the world.

I Learned:

Redo:

Manipulated Variable:

Measured Variable:

Controlled Variable:

(*Teacher note:* According to the U.S. Geological Survey Earthquake Hazards Program, "The magnitude of an earthquake is determined from the logarithm of the amplitude of waves recorded by seismographs. On the Moment Magnitude Scale, magnitude is expressed in whole numbers and decimal fractions. For example, a magnitude 5.3 might be computed for a moderate earthquake, and a strong earthquake might be rated as magnitude 6.3. Because of the logarithmic basis of the scale, each whole number increase in magnitude represents a tenfold increase in measured amplitude; as an estimate of energy, each whole number step in the magnitude scale corresponds to the release of about 31 times more energy than the amount associated with the preceding whole number value.")

Earthquake Monitoring and Mapping Instruction Document

You will use the United States Geological Survey's "Current Worldwide Earthquake List" at *http://earthquake.usgs.gov/regional/neic* to conduct your earthquake monitoring and mapping. On your assigned date, you will use the data presented by the USGS to record one dot on our classroom wall map at the exact latitude and longitude for each earthquake that has measured 5.5 magnitude or higher during the past seven days.

1. Procedure
2. Follow the link on your teacher's website to the National Earthquake Information Center: *http://earthquake.usgs.gov/regional/neic*
3. Click on "Current Worldwide Earthquake List."
4. Click on the Settings icon in the top-right corner of the page.
5. Select "7 Days, Magnitude 4.5+ Worldwide" on the right side of the map.

6. Select "Largest Magnitude First" for the List Sort Order.
7. Remove the checkmark in the box that says, "Only list Earthquakes Shown on Map," so worldwide earthquakes will be shown.
8. You will be plotting dots on our wall map for all Earthquakes with a 5.5 magnitude or greater. Count out how many dots you will need to plot by looking at the listing on the left side of the website map.
9. The location is named for each particular earthquake. Click on each location name, in order to be able to see the location's longitude and latitude (for example, 20.764°N 146.760°E). Mark a dot on the wall map in the classroom for that location.

5A. Shaking Things up Lab

NGSS Alignment

MS-ESS2-2. Construct an explanation based on evidence for how geoscience processes have changed Earth's surface at varying time and spatial scales.

MS-ESS3-2. Analyze and interpret data on natural hazards to forecast future catastrophic events and inform the development of technologies to mitigate their effects.

4-ESS2-2. Analyze and interpret data from maps to describe patterns of Earth's features.

5B.

Mounting Magma Lab

Problem

How do volcanoes differ from each other?

Prediction

Describe, in one sentence, the differences between cinder cone, shield volcanoes, and stratovolcanoes.

(*Teacher note:* Have students share several predictions out loud, so misconceptions can be anticipated and explained.)

Thinking About the Problem

Are all volcanoes alike? Volcanoes form where magma burns through the crust, at subduction (*sub* + *ducere* "to lead under" in Latin) zones, at spreading centers, or at "hot spots" like Hawaii. Volcanoes have two major sections. The *crater* is the pit at the top portion of the volcano. The *vent* is a pipelike structure that connects the underground magma chamber to the crater.

Scientists classify the three types of volcanoes as shield volcanoes, cinder cones, and stratovolcanoes. Shield volcanoes are the largest of all volcano types. They are generally not explosive and are built by the accumulation of lava flows that spread out widely. Lava flows that spread are very fluid, and are described as having low viscosity. *Viscosity* refers to the level of ease with which a fluid flows. Shield volcanoes form broad, gentle slopes, and can be tens of kilometers across, and thousands of meters high. Kilauea and Mauna Loa Volcanoes, in Hawaii, are examples of active shield volcanoes.

Cinder cones are the smallest. They have steep sides formed mainly by the piling up of ash, cinders, and rocks. All of these materials were explosively erupted out of the vent of the volcano, and are called pyroclastic, which means "fire-broken." The cinders shoot out of the volcano like fireworks, and then fall back to the ground. As they fall, they pile up to form a symmetrical, steep-sided cone around the vent. Sunset Crater in Arizona and Paricutin in Mexico are well-known examples of cinder cones.

A stratovolcano is the most common type of volcano on Earth. Also called a composite volcano, it is built up of lava flows layered with pyroclastic material. Scientists believe that the layering represents a history of alternating explosive and quiet eruptions. Young stratovolcanoes are typically steep-sided and symmetrically cone-shaped. There are many active stratovolcanoes in North America, including Mount Saint Helens in Washington. Other well-known volcanoes include Mount Ranier, Mount Shasta, Mount Mazama (Crater Lake), and Redoubt Volcano in Alaska. Mount Fuji in Japan and Mount Vesuvius in Italy are other famous stratovolcanoes.

Recently, researchers have started to find patterns in volcanic activity related to the changes in the speed of Earth's rotation. This difference is only on the order of milliseconds, but it may be causing significant stresses and pressures inside our planet due to the energy required to alter the spin. Those stresses and pressures may be making it easier for the magma to rise to the Earth's surface.

In this lab, you will put together a model of a volcano. Although volcanoes erupt due to built-up geologic pressures, you will use a chemical reaction to simulate those pressures in the eruption of a model volcano.

Write three main points from the "Thinking About the Problem" reading:

1. A stratovolcano is ...

2.

3.

Materials

- Film canister (one per lab group)
- pH paper
- Scissors
- Two Alka-Seltzer tablets
- Volcano pattern (one per student)
- Water (both warm and cold)

DRAW LABELED SKETCH OF EXPERIMENTAL MATERIALS HERE.

Procedure

1. Cut out the volcano pattern (Figure 5.2, p. 199), along the outer lines.
2. Fold the model volcano along the inner lines, and then tape the tabs together.
3. Place the empty film canister in the center of the cone.
4. Cut each Alka-Seltzer tablet in half, enabling you to conduct repeated trials. Place one piece and one drop of red food coloring into the film canister.
5. Quickly add some cold water (testing for pH) to the canister and watch the mounting magma.
6. Use your lab partner's volcano model, and repeat step 4. Then quickly add some warm water (testing for pH) to the canister and watch.
7. Study the results and along with determining the effect of water temperature on reaction rates, also determine which type of volcano your model represents.
8. Draw a labeled sketch of your experimental materials in the box above.

Analysis

1. Which liquid, warm water or cold water, produced the best simulation of a volcanic eruption, and why?
2. What results did you see when you tested the liquids (before and after) for pH?
3. After studying the results, which type of volcano has been produced by you?
4. Make three detailed observations about your results from this experiment.
5. Compare and contrast your results with those of the other lab groups. Describe what you notice.
6. (Enrichment) Investigate what types of rocks are produced by each of the types of volcanoes mentioned in this lab.
   - Cinder Cone:
   - Shield Volcanoes:
   - Stratovolcanoes:
7. (Enrichment) Name and describe one planet or moon in our solar system that shows evidence of volcanic activity. Describe the evidence, as well.
8. (Enrichment) Create a short *Keynote* presentation on the history of eruptions from Mount St. Helens. Contrast it with the forecast of eruptions of Mount Rainier.
9. (Enrichment) Investigate the rationales and issues associated with people who chose to live at the foot of a volcano. Design a possible plan for solutions to these problems.
10. (Enrichment) Using the map that you developed for *Shaking Things up*, plot all of the recently active volcanoes on Earth. What patterns do you notice?

Learning Target

Construct and test a model of a volcano.

I Learned:

Redo:

Manipulated Variable:

Measured Variable:

Controlled Variable:

FIGURE 5.2.

VOLCANO PATTERN

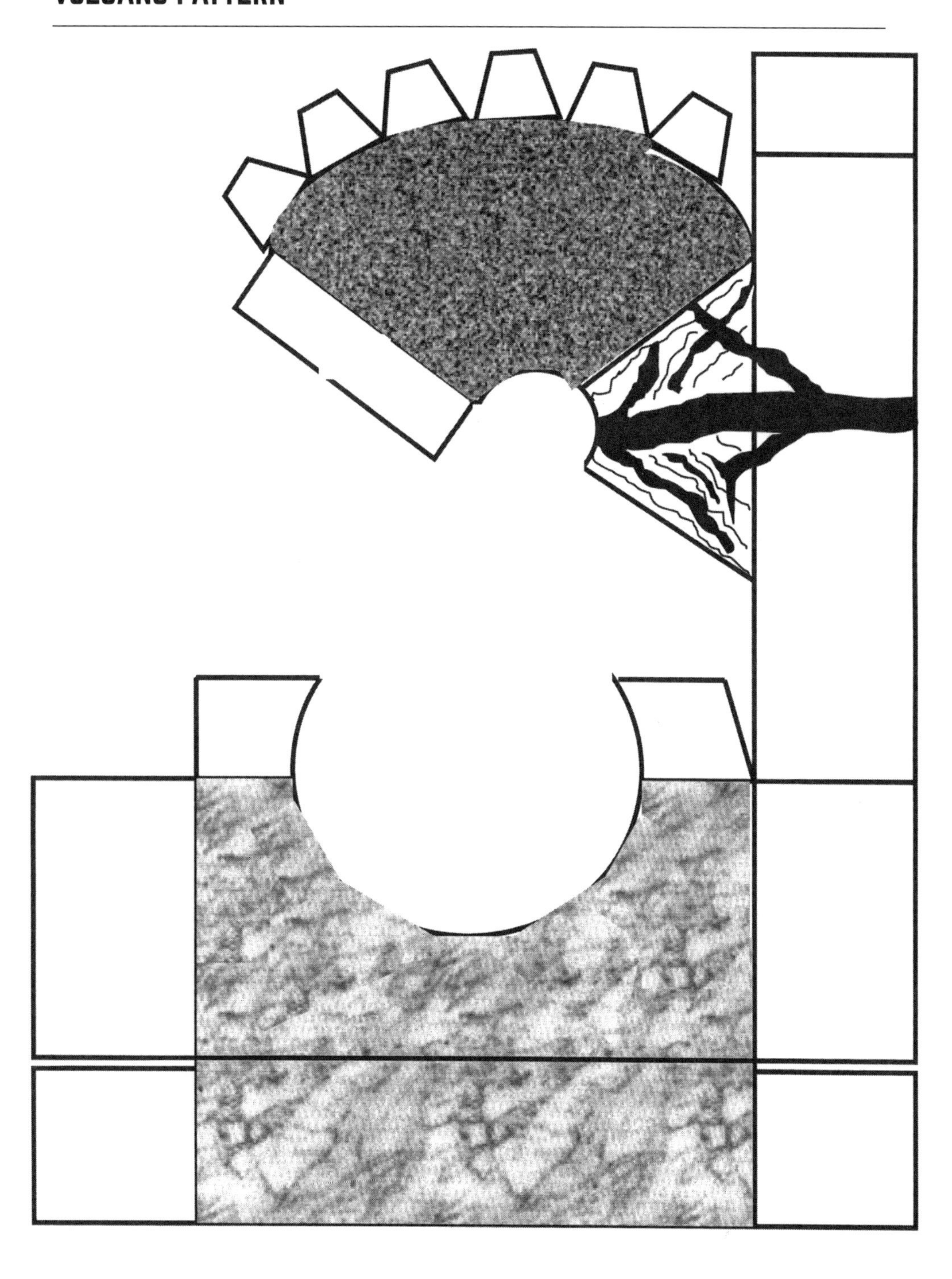

5B. Mounting Magma Lab

NGSS Alignment

MS-ESS1-4. Construct a scientific explanation based on evidence from rock strata for how the geologic time scale is used to organize Earth's 4.6-billion-year-old history.

MS-ESS2-2. Construct an explanation based on evidence for how geoscience processes have changed Earth's surface at varying time and spatial scales.

MS-ESS3-2. Analyze and interpret data on natural hazards to forecast future catastrophic events and inform the development of technologies to mitigate their effects.

5C. Hypothesizing About Plates Activity

Clock Hour Appointments

Students form a partnership with five different people in the classroom. When forming a partnership, both partners must have identical appointment times scheduled for each other (Figure 5.3).

FIGURE 5.3.

CLOCK HOUR DIAGRAM

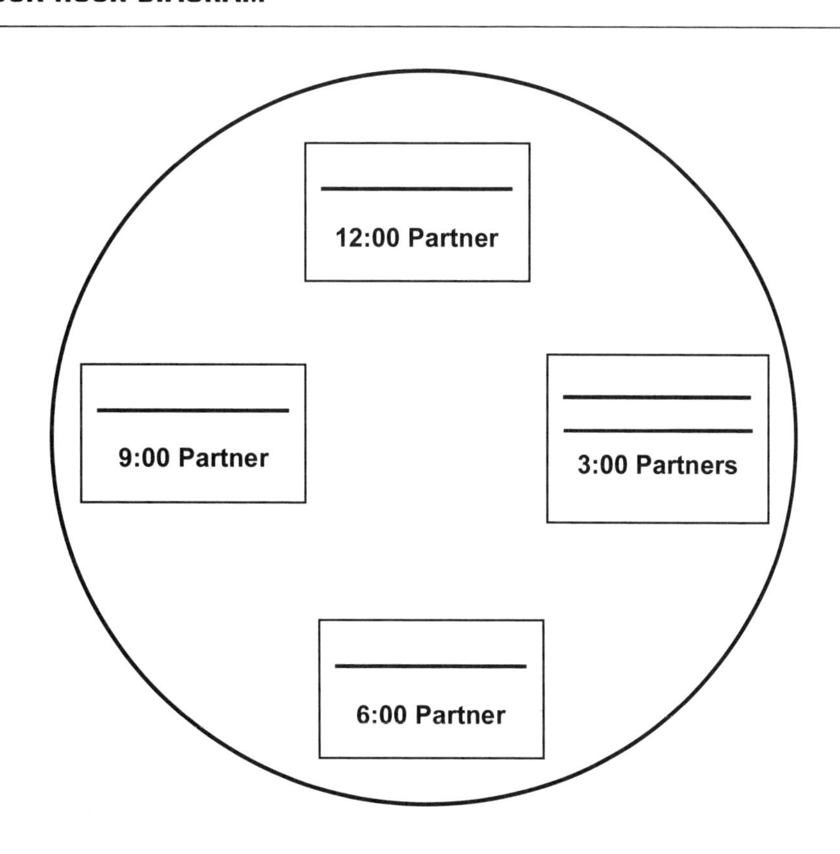

The Language of Science

(To be read aloud): Scientists use their words carefully in order to be precise. For example, *theory*, *law*, and *hypothesis* all have very different and agreed-upon meanings. All of the meanings are based on the differences between claims, evidence, and reasons. Outside of science, you might misuse a term by saying something is "just a theory," meaning it may or may not be true. In science, however, a theory is an explanation that needs to be continuously supported by observable, measurable evidence and experimentation.

Part A

Do a read-aloud with your two 3:00 partners, and then return to your seats. In the read-aloud, one student reads, and then shares two main points. The other students listen, and then each uses a "paraphrase starter" (Figure 5.4) to summarize a main point.

FIGURE 5.4.

PARAPHRASE STARTERS

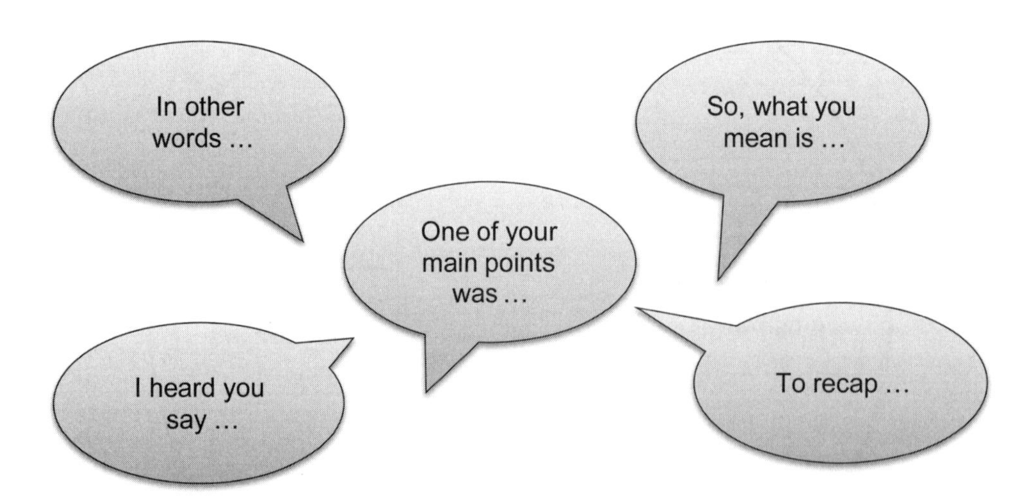

Hypothesis

(Partner #1 reads and shares, while #2 and #3 each use a "paraphrase starter.")

A hypothesis is an explanation based on your observations, which can be tested by experiment. It is not a prediction (an educated guess), but it might be based on a prediction. Sometimes it helps to think of the hypothesis as an "if/then prediction".

Scientists make hypotheses based on what they observe about the world around them.

A prediction that a scientist makes, explaining a natural event, is sometimes called an "educated guess." For example, "I predict it will be blue." This is different from a hypothesis, which is an explanation that can be tested, for example, "It will be blue because blue is higher in energy."

Usually, a hypothesis can be supported or refuted (proven to be false) through experimentation or more observation. Example: If you see no difference in the cleaning ability of various dishwashing detergents, you might hypothesize that cleaning effectiveness is not affected by which detergent you use. You can see this hypothesis can be disproven if you can remove a food particle with one detergent, but not the other. However, you cannot prove the hypothesis. Even if you never see a difference in the cleanliness of your dishes after trying many detergents, there might be one you haven't tried that could be different.

Theory

(Partner #2 reads and shares, while #1 and #3 each use a "paraphrase starter.")

A scientific theory summarizes several hypotheses that have been supported throughout measurable and observable experiments. A theory can be formulated only after evidence accumulates enough to support all of the associated hypotheses. Theories can be proven wrong. However, theories will be valid as long as there is no measurable and observable evidence to refute them.

A scientific theory begins as one or more untested ideas about why something happens (this is when we consider it to be a hypothesis). To become a theory, an idea must be thoroughly tested. It becomes an accurate and predictive description of the natural world. It can also be revised to include brand new evidence that is presented.

Example: It is known that on February 15, 2013 in Chelyabinsk, Russia, there was an explosion equivalent to the detonation of about 500 kilotons of TNT (which is about 30 times more energy than was released from the atomic bomb detonated in Hiroshima, Japan). Many hypotheses have been proposed for what caused the explosion and the resulting injuries to 1,500 people. It is theorized that a meteor, from a near-Earth asteroid, caused the explosion. Is this theory a fact? No. The

event is a recorded fact, and has observable evidence to back it up. Is this theory generally accepted to be true, based on evidence? Yes. Can this theory be shown to be false and need changing? Potentially.

Law

(Partner #3 reads and shares, while #1 and #2 each use a "paraphrase starter.")

A law, in the language of science, describes what things will do. A law is an analytic statement, often based on a proven formula. Scientific laws let us calculate quite a bit about how events will occur. They help us explain things, but not describe them fully. One way to tell a law and a theory apart is to ask if the description gives you the ability to explain "why." For example, Newton's law of universal gravitation tells us that "Every point mass attracts every single point mass by a force pointing along the line intersecting both points. The force is directly proportional to the product of the two masses and inversely proportional to the square of the distance between the point masses." Newton could use this law to predict the behavior of a dropped object, but he couldn't explain why it happened.

We can use Newton's law of gravity to calculate how strong the gravitational pull is between the planets and the Sun, or between the Earth and the Moon, or between the Earth and any object you drop. This allows us to calculate many things, such as acceleration, how long the planet will take to orbit, how fast it would be going during its revolution, how much energy it will take to pick an object up, and so on.

If you're asked to define *hypothesis*, *theory*, and *law*, it is important is to realize they don't all mean the same thing and cannot be used interchangeably.

Part B

Use your 6:00 partner to read the directions below and then together accurately complete the 10 labeling tasks, with justifications.

Directions: Below are simple statements. The complicated part involves labeling each of them as a *hypothesis*, *theory*, or *law* and explaining why. The justification you give could be a way to disprove it, an argument for or against it, or a way to explain "why."

1. The planets are held in orbit around the Sun by its gravitational pull on them. (*Teacher note:* Answer is *theory*)

   Circle the Best Label: Hypothesis Theory Law

   Justification for Decision:

2. There are always coins between the seats in cars. (*Teacher note:* Answer is *hypothesis*)

   Circle the Best Label: Hypothesis Theory Law

   Justification for Decision:

3. Hot air rises because cold air falls. (*Teacher note:* Answer is *theory*)

   Circle the Best Label: Hypothesis Theory Law

   Justification for Decision:

4. Every time I forget to turn the volume off, my phone will ring. (*Teacher note:* Answer is *hypothesis*)

   Circle the Best Label: Hypothesis Theory Law

   Justification for Decision:

5. Matter can be neither created nor destroyed; it only changes form. (*Teacher note:* Answer is *law*)

   Circle the Best Label: Hypothesis Theory Law

   Justification for Decision:

6. Volcanoes in ancient times behaved like volcanoes today. (*Teacher note:* Answer is *theory*)

   Circle the Best Label: Hypothesis Theory Law

   Justification for Decision:

7. All of the energy flowing, and matter cycling, within and among the Earth's systems is derived from the Sun and Earth's hot interior. (*Teacher note:* Answer is *law*)

   Circle the Best Label: Hypothesis Theory Law

   Justification for Decision:

8. If the barometric pressure decreases, then clouds will form. (*Teacher note:* Answer is *hypothesis*)

   Circle the Best Label: Hypothesis Theory Law

   Justification for Decision:

9. Landscapes change over time due to geoscience processes. (*Teacher note:* Answer is *theory*)

   Circle the Best Label: Hypothesis Theory Law

   Justification for Decision:

10. Mapping the history of natural hazards in a region, and developing an understanding of geologic forces involved, can help forecast the locations and likelihoods of future events. (*Teacher note:* Answer is *law*)

    Circle the Best Label: Hypothesis Theory Law

    Justification for Decision:

Part C

Working with your 9:00 partner, briefly share what each of you wrote in Part B above. Discuss your answers to each, looking closely at any similarities and differences.

1. Describe one similarity in your answers.
2. Describe one difference in your answers.
3. Change any answers, together, that you now feel should be changed.

Part D

Read the paragraphs below with your 12:00 partner, underline three main points, and then write two "I learned …" statements.

Alfred Wegener, a brilliant German interdisciplinary scientist (meteorologist, polar researcher, and geophysicist) who lived from 1880 to 1930, said,

> Scientists still do not appear to understand sufficiently that all earth sciences must contribute evidence toward unveiling the state of our planet in earlier times, and that the truth of the matter can only be reached by combing all this evidence. It is only by combing the information furnished by all the earth sciences that we can hope to determine "truth" here, that is to say, to find the picture that sets out all the known facts in the best arrangement and that therefore has the highest degree of probability. Further, we have to be prepared always for the possibility that each new discovery, no matter what science furnishes it, may modify the conclusions we draw. (*The Origins of Continents and Oceans*, 1915)

Alfred Wegener studied all of types of evidence and observations, and then proposed the theory of plate tectonics. As a careful observer, he noticed that plant fossils of the late Paleozoic age were found on several different continents, but were nearly identical to each other. This suggested that the plants may have evolved together on a single large land mass. Wegener was especially intrigued by the occurrences of plant and animal fossils found on the matching coastlines of South America and Africa, which are now widely separated by the Atlantic Ocean. To Wegener, the presence of identical fossil species along these coastal parts was the most compelling evidence that the continents were once joined.

That land mass is now referred to as Pangaea, and the theory is that over the past 300 million years, the continents separated and shifted toward what we observe today.

Why are dinosaur fossils found in Antarctica? This provides yet more evidence for continental shifting due to plate tectonics. The northern continents with which we are familiar were all at the equator 300 million years ago. The discovery of fossils of tropical plants in Antarctica led to the conclusion that it must have, at one time, been situated in a more temperate climate where tropical vegetation could grow.

The theory of plate tectonics was one of the most important and revolutionary geological theories of all time. It took decades after being first proposed to become generally accepted by scientists, however. Steadily accumulating evidence finally prompted its acceptance, and its eventual elevation to the status of *theory*.

5C. Hypothesizing About Plates Activity

NGSS Alignment

MS-ESS1-2. Develop and use a model to describe the role of gravity in the motions within galaxies and the solar system.

MS-ESS2-1. Develop a model to describe the cycling of Earth's materials and the flow of energy that drives this process.

MS-ESS2-2. Construct an explanation based on evidence for how geoscience processes have changed Earth's surface at varying time and spatial scales.

MS-ESS2-3. Analyze and interpret data on the distribution of fossils and rocks, continental shapes, and seafloor structures to provide evidence of the past plate motions.

MS-ESS2-5. Collect data to provide evidence for how the motions and complex interactions of air masses results in changes in weather conditions.

MS-ESS3-2. Analyze and interpret data on natural hazards to forecast future catastrophic events and inform the development of technologies to mitigate their effects.

5D.

Cracking up With Landforms Lab and Landforms Formative Assessment

Problem

How has the movement of tectonic plates caused geologic changes on Earth?

Prediction

Answer the problem statement above.

(*Teacher note:* Have students share several predictions out loud, so misconceptions can be anticipated and explained.)

Thinking About the Problem

Our Earth is made of several layers—crust, mantle, outer core, and inner core. The crust and the upper part of the mantle make up the lithosphere. The lithosphere is made up of tectonic plates, which rest on the hot upper mantle. Take a moment to find a diagram of Earth's layers, using your tablet. Main point #1 below must be about what you observed in this Earth layers diagram.

The theory of plate tectonics states that the crust of the Earth is composed of seven major plates and numerous smaller plates. This theory also says that the continents have changed position over time. Gravity and convection currents in the mantle move tectonic plates over Earth's surface. As the plates move, they interact at plate boundaries. At plate boundaries, plates may converge (come together), diverge (move apart) or slip past each other in a horizontal motion (transform boundary).

Now, use your tablet to find a diagram that names the seven major tectonic plates on Earth. Go back to the world map you used for the Shaking Things Up Lab. Name any major tectonic plates for which you have found evidence. Main point #2 below must be about what you can now observe on your map.

New crust is formed at *divergent boundaries*. Features include mid-ocean ridges and rift valleys. Crust is destroyed at *convergent boundaries. Subduction boundaries* form island arcs, deep-ocean trenches, and coastal mountains. *Collision boundaries* can form mountains. At *transform boundaries*, as two plates scrape past one another, the Earth's crust breaks due to friction and earthquakes may occur. Finally, take a moment to research on the internet. Find a diagram that shows the motions involved with divergent, convergent, and transform plates, and examine it closely. Main point #3 below must be about the motions in the diagram you found.

In this lab, a wheat cracker represents the thinner, denser oceanic crust. The Goldfish cracker represents the thicker, less dense continental crust. The frosting represents lava or magma that the tectonic plates are floating on.

Write three main points from the "Thinking About the Problem" reading:

1.

2.

3.

Materials

- 1 box of wheat crackers per class
- 1 box of Goldfish crackers per class
- ½ cup of frosting per group
- 1 piece of wax paper per group
- 1 plastic knife per group

Procedures for Day 1

Part 1: Divergent Plate Boundaries (Oceanic vs. Oceanic)

1. Everyone in the group must wash their hands with soap and water.
2. Do not eat any of the crackers, treats, or frosting.
3. Set out your piece of wax paper and gather your materials.
4. Obtain two square wheat crackers.
5. Using the plastic knife, spread a thick layer of frosting in the center of the wax paper. It should be approximately 10 cm square, and about 2 cm thick.
6. Lay the two wheat crackers side by side on top of the frosting so they are touching (like dominoes, side-by-side).
7. To simulate the movement of diverging oceanic plates, firmly push down on the crackers (without breaking them) while slowly moving them apart. Move them about 2 cm apart.
8. Take three pictures with your iPad (before, during, and after) and insert them in the first box in Data Table 5.1 (p. 213).

Part 2: Converging Plate Boundaries (Oceanic vs. Continental)

1. Remove both wheat crackers from the frosting (from Part 1) and set them aside.
2. Use the plastic knife to respread the layer of frosting in the center of the wax paper. Add to the frosting, if necessary. It should cover an area that is approximately 10 cm square, and about 2 cm thick.
3. Lay one wheat cracker and one Goldfish cracker end-to-end on top of the frosting. The wheat cracker represents the thinner, denser ocean plate. The Goldfish cracker represents the thicker, less-dense continental plate.
4. Simulate the movement of a converging oceanic plate with a continental plate. Press down firmly (without breaking) a bit harder on the wheat cracker than the Goldfish cracker, to represent the greater density of oceanic plate. The ocean plate (wheat cracker) should "subduct" underneath the continental plate (Goldfish cracker).
5. Take three pictures with your iPad (before, during, and after) and insert them in the second box in Data Table 5.1 (p. 213).

Part 3: Converging Plate Boundaries-Continental vs. Continental

1. Use the knife to respread the same frosting in the center of the wax paper. Next, get two new Goldfish crackers, in order to have two continental plates.
2. Put the two Goldfish crackers onto your bed of frosting. Push the two Goldfish crackers together to simulate a convergent plate boundary. Use enough force to fold and deform the meeting edges of the "plates."
3. Take three pictures with your iPad (before, during, and after) and insert them in the third box in Data Table 5.1.

Part 4: Transform Plate Boundaries

1. Respread your frosting one last time with the plastic knife. Add to the frosting, if necessary, from your original amount. It should cover an area that is approximately 10 cm square, and about 2 cm thick.
2. Obtain two new wheat crackers and place the two side by side on top of the frosting.
3. Simulate the movement along a transform boundary. This will be a bit tricky—you'll have to push the wheat crackers together with moderate force while sliding one away from you and pulling the other one toward you. Start with a very light sliding pressure and keep increasing until the crackers slide past each other. If you are doing this correctly, the crackers should break.
4. Take three pictures with your iPad (before, during, and after) and insert them in the fourth box in Data Table 5.1.

Cleanup

1. Put your plastic knife, wax paper, and all uneaten materials in the garbage.
2. Wipe up your table with a wet paper towel. Make it spotless.
3. Wash your hands and sit down in your desks.

DATA TABLE 5.1.

LABELED IMAGES OF PLATE BOUNDARY MODELS

Part 1: Divergent Plate Boundaries (Oceanic vs. Oceanic)	**Part 2: Convergent Plate Boundaries (Oceanic vs. Continental)**
Part 3: Converging Plate Boundaries (Continental vs. Continental)	**Part 4: Transform Plate Boundaries**

Procedures for Day 2

Part 1: Divergent Plate Boundaries (Oceanic vs. Oceanic)

Look at the Part 1 pictures to answer the following questions (and research on the internet as well).

1. What happened to the frosting between the crackers?
2. What do the wheat crackers represent?
3. What does the frosting represent?
4. Provide an example of a location where this type of boundary is found on Earth.
5. What type of feature would be produced when two continental plates move apart like this?
6. What type of feature would be produced when two oceanic plates move apart like this?

Part 2: Convergent Plate Boundaries (Oceanic vs. Continental)

Look at the Part 2 pictures to answer the following questions (and research on the internet, as well).

1. Explain why the wheat cracker and Goldfish cracker represent the type of crust they do.
2. What happens when the wheat cracker and Goldfish cracker meet?
3. What term is used to describe one plate moving beneath another?
4. Provide an example of a location where this type of boundary is found on Earth.
5. What happens to a real tectonic plate being subducted?
6. What features are formed along a subduction zone?

Part 3: Converging Plate Boundaries (Continental vs. Continental)

Look at the Part 3 pictures to answer the following questions (and research on the internet as well).

1. Explain what happens to the edges of the Goldfish crackers.
2. What does the deformation of the Goldfish crackers represent in real life?
3. Provide an example of a location where this type of boundary is found on Earth.

Part 4: Transform Plate Boundaries

Look at the Part 4 pictures to answer the following questions (and research on the internet as well).

1. Provide an example of a location where this type of boundary is found on Earth.
2. Why don't the crackers move easily at first?
3. What real geologic event is simulated when the crackers finally move past each other and breakage occurs?
4. Nothing happens to the crackers in the beginning, but as the pressure is increased, the crackers finally break. Explain how this is similar to the situation along the San Andreas Fault in California.

Learning Target

Understand how the movement of tectonic plates causes geologic changes on Earth.

Helpful Video

http://geology-guy.com/teaching/iac/animations/terrane_formation.htm

Sample Landforms Formative Assessment Learning Target

Describe how Minnesota's landscape has been shaped by geological events, such as glaciers, volcanoes, and oceans. For this formative assessment, the author collected six images (Figures 5.5–5.10) showing three different landforms in the state of Minnesota, along with three images of the various geologic events that caused these particular landforms to be generated. Students are asked to match the landforms (numbered 1, 2, and 3) with the correct images of volcano, glacier, or moving water (labeled A, B, and C) that formed them. (*Teacher note:* In order to see this document, with the assessment questions on it, students may view it from your Google website and then respond to the Google form, or they can respond to the questions using a document uploaded to *Schoology*. With iPads, students are able to four-finger-swipe between two documents, if needed. Answers are in parentheses.)

____ 1. FIGURE 5.5.

BASALT FORMATIONS IN HAWAII (PHOTO FROM MARCY OATES) (C)

____ 2. FIGURE 5.6.

NISQUALLY RIVER IN THE CASCADE MOUNTAIN RANGE (A)

____ 3. FIGURE 5.7.

SANDSTONE ROCK FORMATIONS IN COLORADO SPRINGS (B)

A. FIGURE 5.8.

GLACIERS ON MOUNT RAINIER IN WASHINGTON

B. FIGURE 5.9.

MOVING WATER IN LAKE SUPERIOR

C. FIGURE 5.10.

MOUNT SAINT HELENS VOLCANO IN WASHINGTON

5D. Cracking up With Landforms Lab and Landforms Formative Assessment

NGSS Alignment

MS-ESS2-2. Construct an explanation based on evidence for how geoscience processes have changed Earth's surface at varying time and spatial scales.

MS-ESS2-3. Analyze and interpret data on the distribution of fossils and rocks, continental shapes, and seafloor structures to provide evidence of the past plate motions.

MS-ESS3-1. Construct a scientific explanation based on evidence for how the uneven distributions of Earth's mineral, energy, and groundwater resources are the result of past and current geoscience processes.

6

EARTH'S WEATHER

6A.

Wondering About Water Lab

Problem

How does the hydrologic cycle affect the world around me?

Prediction

Describe, in one sentence, the answer to the problem statement.

(*Teacher note:* Have students share several predictions out loud, so misconceptions can be anticipated and explained.)

Thinking About the Problem

Where does the water in rain come from and where does it go? Our seemingly inexhaustible supply of water is actually used over and over again. Water moves from the ocean to air and land and from the land to ocean and air. This continuous movement of water in a cyclic pattern is called the hydrologic (*hydro* "water" in Greek) cycle. And it is very important in the science of meteorology.

Water molecules with the highest amount of energy are in the form of gases, called water vapor. If these molecules give off their energy and cool in temperature, then they change phase into liquid water. Water molecules in the liquid phase move around more slowly than gas molecules. If the liquid water molecules shed even more energy and their temperature and motion drops low enough, they become solid ice.

In nature, the Sun's energy warms the water in the oceans. This tremendous amount of energy causes the surface water molecules to change phase and evaporate as water vapor. As the water vapor rises in the atmosphere (atmos "vapor" in Greek), it cools and condenses into liquid droplets. Most of these droplets continue cooling to ice crystals and snowflakes. Evaporation continues as the droplets and crystals grow in size, until they eventually fall back to Earth. Precipitation falling on the Earth's surface is dependent on the temperature changes below the clouds. The precipitation collects on the land surface and may flow back to the oceans, completing the cycle.

Most of the Earth's total amount of water is contained in the oceans, a volume estimated at 1,350,000,000 km^3. Other reservoirs, for example glaciers (27,500,000 km^3), and groundwater (8,200,000 km^3), lakes and streams (206,700 km^3), hold significant water as well. Although it is in a continuous state of change due to evaporation and precipitation, our atmosphere is estimated to hold 13,000 km^3 of water.

In this lab, you will also read a story, "Water Wonders," to develop ideas of your own about the hydrologic cycle steps (see sample in Figure 6.1 below).

Write three main points from the "Thinking About the Problem" reading:

1.

2.

3.

FIGURE 6.1.

LABELED SKETCH OF HYDROLOGIC CYCLE

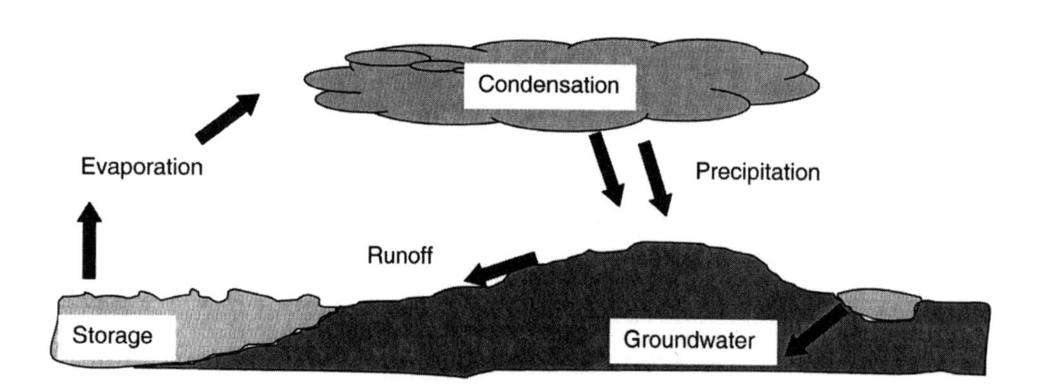

Analysis

1. Read "Water Wonders" (Figure 6.2, p. 224).
2. Use this story to help give you new ideas in describing the journey of your very own water molecule through any three steps of the hydrologic cycle.
3. Share your water molecule's journey in a five-scene comic book (Figure 6.3, p. 225).
4. (Enrichment) Investigate what types of engineering devices are used to produce drinking water from ocean water. Design and draw a labeled diagram of a desalination device made from household items, which could work to provide fresh safe drinking water.
5. (Enrichment) Research the percentage of time that a typical water molecule would spend in each particular step of the hydrologic cycle. For example, if given 100 days, how many days would the water molecule spend in the ocean, in the atmosphere, in a lake, or in a cloud?

FIGURE 6.2.

WATER WONDERS

Once upon a time, a royal family lived in a palace in the clouds. Two sisters, both water molecule princesses, were always looking for adventures. Their names were Maureen and Mary Molecule, and they were both kind-hearted, wonderful water molecules. For fun, one day, they fell to Earth as part of a raindrop with many other molecules from their kingdom.

Maureen and Mary landed in a bur oak tree, rolled over an acorn, and then dripped gently to the ground. Both had many good friends with them—molecules are so very small that it took lots and lots of molecules to make up the tiny raindrop that Maureen and Mary were in.

A short time after they landed on the ground, the sun came out. The sun warmed the ground and all of the water molecules on it. Maureen and Mary started to feel wonderfully warm. They evaporated back into the atmosphere and eventually continued their adventures around Earth as part of the water molecule cycle.

During Maureen and Mary Molecule's many travels in the water cycle they spent time stored on the surface of the Earth in glaciers, melting and flowing into lakes, relaxing underground among rocks and soil, and even traveling through living things. They were able to join other water molecules to become a water droplet in a cumulus cloud, freezing and falling to Earth as a snowflake. Eventually Maureen and Mary could flow to the oceans, be transpired by plants, or be evaporated directly back up into their palace in the sky.

Throughout all of Maureen and Mary Molecule's journeys, the amount of water on Earth and in our atmosphere remains the same—it just changes from solid to liquid to gas, and does some traveling. You see, water molecules are pretty tough, and it is very hard to hurt them. Maureen and Mary are strong. As water molecules, they are made up of three atoms: two small hydrogen atoms (H_2), and one larger oxygen atom (O), tightly connected. Water molecules (H_2O), like Maureen and Mary, don't change their appearance when they are cooled, warmed, or under pressure. In ice, water, and water vapor, water molecules stay the same.

For the analysis section, you need to use your creativity to describe the journey of any water molecule through any three steps of the water cycle. You need to make your description into a five-scene comic strip.

FIGURE 6.3.

WATER WONDERS COMIC BOOK

(The *ComicBook!* app is used on the iPad.)

<table>
<tr><th colspan="2">Title:</th></tr>
<tr><td>Page 1: Introduce your main character(s)

Narration</td><td>Page 2: Take your character(s) through any first step of the hydrologic cycle.

Narration</td></tr>
<tr><td>Page 3: Take your character(s) through any next step of the hydrologic cycle.

Narration</td><td>Page 4: Take your character(s) through any third step of the hydrologic cycle.

Narration</td></tr>
<tr><td colspan="2">Page 5: Wrap up your comic book story, showing a final outcome for your comic book character(s).

Narration</td></tr>
</table>

6A. Wondering About Water Lab

NGSS Alignment

MS-ESS2-4. Develop a model to describe the cycling of water through Earth's systems driven by energy from the Sun and the force of gravity.

MS-ESS3-5. Ask questions to clarify evidence of the factors that have caused the rise in global temperatures over the past century.

5-PS2-1. Support an argument that the gravitational force exerted by Earth on objects is directed down.

5-ESS2-1. Develop a model using an example to describe ways the geosphere, biosphere, hydrosphere, and/or atmosphere interact.

5-ESS2-2. Describe and graph the amounts and percentages of water and fresh water in various reservoirs to provide evidence about the distribution of water on Earth.

6B.

Piling up the Water Lab

Problem

What is so special about water?

Prediction

Give a working definition of water molecule.

(*Teacher note:* Have students share several predictions out loud, so misconceptions can be anticipated and explained.)

Thinking About the Problem

What does H_2O mean? Each molecule (*molecula* "small bit" in Latin) of water is made of two hydrogen atoms (H_2) and one oxygen atom (O). What is special about water molecules is that they tend to "stick" to each other (cohesion) and to other molecules (adhesion). They do this because water is built like a magnet, with a positive end and a negative end. This helps it bond well.

Water makes life on Earth possible. It covers almost three-fourths of the surface of our planet. Because there is so much of it, water may seem very ordinary to us, and yet it is unique when compared to all other substances. For example, water is the only substance on Earth that occurs naturally in all three states—solid, liquid, and gas. In addition, solid H_2O (ice) is less dense than its liquid form (water), so it floats. Most other solids are denser than their liquid form, so they sink! Another difference, with respect to water, is that large amounts of energy must be added to water to achieve even a relatively small change in temperature. That is why our oceans moderate the temperatures of coastal communities on Earth.

Write three main points from the "Thinking About the Problem" reading:

1.

2.

3.

Materials

- 8 oz. drinking glass
- Dish soap
- Eyedropper
- Variety of water containers (assortment of five glasses, buckets, or bowls)
- Pennies (or similar replacement item; control for size)
- Other coins

Procedure

1. Predict which of your five large containers (each full to the rim with water) will be able to withstand the addition of the greatest number of pennies (or replacement item) without spilling over. Test and record your results in Data Table 6.1 (p. 230). Take photos with your iPad while conducting this step.
2. Place a dry penny on a piece of paper towel.
3. Predict the number of drops you can pile on the penny before water runs over the edge.
4. Test and record for each particular coin in Data Table 6.2 (p. 230). Take photos with your iPad during this step.
5. Take a photo for your labeled image (box on p. 229) of the water on the surface of the coin just before the water spilled over.
6. Conduct the same tests with the soapy water and record your results in Data Table 6.3 (p. 230).

Analysis

1. Describe the shape of the water as it "sits" on a coin.
2. Why does water pile up on a coin, rather than spilling over the edges immediately? How is the soapy water different? (Describe the science behind your thoughts. Review the "Thinking About the Problem" section of this lab.)
3. Use science concepts to suggest reasons why each of the five containers holds a different number of pennies. See student sample, Figure 6.4, p. 232.
4. Explain "surface tension" as if you were explaining it to a second grader.
5. (Enrichment) Investigate the difference between the surface tension of tap water and salt water. Although the addition of impurities, such as salt, decreases the cohesion between water molecules, it also increases the density of water (allowing things to float more easily). Draw a Comparing and Contrasting Diagram to show your findings.
6. (Enrichment) Prepare an *Explain Everything* video that describes either "capillary action" (the mechanism by which plants transport water from their roots throughout the plant) or "water striders" (animals that take advantage of water's surface tension to live on the surface of the water).

INSERT LABELED IMAGE FOR PROCEDURE #5 HERE.

DATA TABLE 6.1.

PREDICTIONS AND RESULTS FOR FULL CONTAINERS

CONTAINER #	DESCRIPTION	PREDICTED # OF PENNIES	ACTUAL # OF PENNIES
1			
2			
3			
4			
5			

DATA TABLE 6.2.

DROPS OF WATER ON COINS

COIN	PREDICTED # OF DROPS	ACTUAL # OF DROPS
Penny		
Nickel		
Dime		
Quarter		

DATA TABLE 6.3.

DROPS OF SOAPY WATER ON COINS

COIN	PREDICTED # OF DROPS	ACTUAL # OF DROPS
Penny		
Nickel		
Dime		
Quarter		

Learning Target

Develop an understanding about how water is unique when compared to all other substances.

I Learned:

Redo:

Manipulated Variable:

Measured Variable:

Controlled Variable:

Liter Bottle World Water Analogy

1. Show a 1-liter bottle, full of water. It represents all of the water on Earth.
2. Pour 30 ml into a graduated cylinder. This represents all of the freshwater.
3. Pour off 6 ml into smaller graduated cylinder. The remaining 24 ml represents what is stored in glaciers, which will eventually melt into and dilute oceans. The 6 ml represents nonfrozen rivers, lakes, and aquifers.
4. Pour off 1.5 ml of the 6 ml, into a small graduated cylinder. This represents groundwater.
5. Using an eyedropper, take one drop of that 1.5 ml. This represents the accessible, unpolluted ground water.
6. Remind students that many people in the United States use 50% of that one drop to water their lawns.

FIGURE 6.4.

STUDENT SAMPLE OF PILING UP THE WATER FROM ELECTRONIC SCIENCE NOTEBOOK

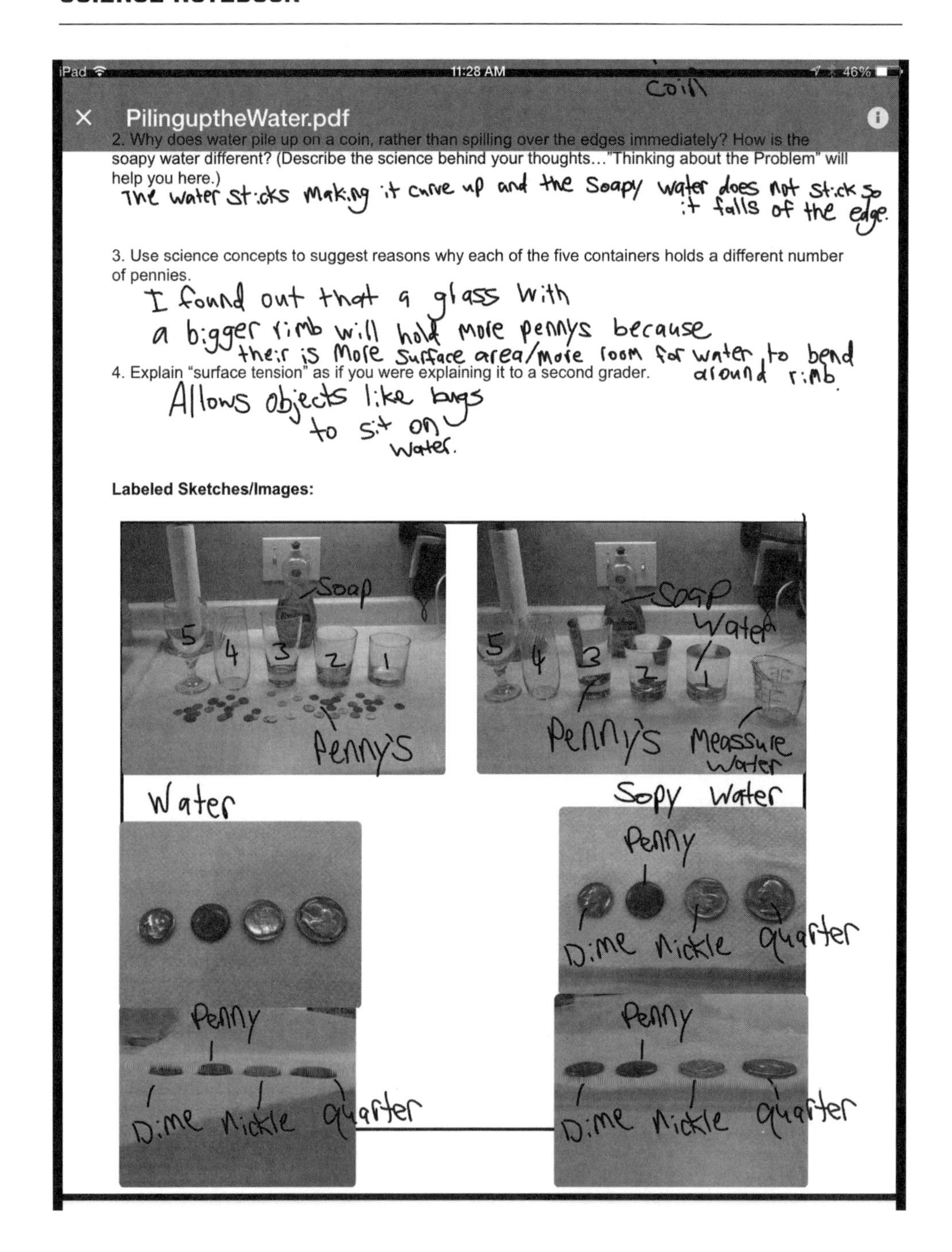

6B. Piling up the Water Lab

NGSS Alignment

MS-ESS2-4. Develop a model to describe the cycling of water through Earth's systems driven by energy from the Sun and the force of gravity.

MS-PS1-1. Develop models to describe the atomic composition of simple molecules and extended structures.

6C.

Phasing in Changes Lab

Problem

What are the two things that can happen when heat is added to water?

Prediction

Answer the problem statement above.

(*Teacher note:* Have students share several predictions out loud, so misconceptions can be anticipated and explained.)

Thinking About the Problem

What happens when you heat water? The first is that the temperature of the water rises, or increases. The second is that a "change of phase" may occur.

A phase change refers to whether a substance like water, is a solid (ice), liquid (water), or a gas (water vapor). *Freezing* refers to a change from liquid to solid. *Melting* refers to a change from solid to liquid. *Boiling* refers to a change from liquid to gas. *Condensing* refers to a change from gas to liquid. Sometimes a substance, such as carbon dioxide (its solid form is called *dry ice*, because it does not melt into a liquid), can skip a step. *Sublimation* refers to these cases, when a substance goes directly from solid to gas.

In each phase change, the spacing between the molecules that make up the substance change. In a solid, the molecules are packed very closely together. The molecules are still moving, vibrating back and forth. In a liquid, the space between molecules increases to the point that they can flow around each other. In a gas, the molecules dart around freely, occasionally colliding with each other.

A graph of your data from this lab will show that temperatures stop increasing when phase changes are occurring. Once the molecules have separated to the distance needed for the substance to become a different phase, and then the temperature begins to measure this new molecular motion.

Write three main points from the "Thinking About the Problem" reading:

1. The molecules in a solid are ...

2.

3.

Procedure

1. No paper or writing utensils allowed at lab stations. Concentrate on safety first, please.
2. Create two data tables, giving them descriptive titles.
3. One should show temperature versus time for water boiling.
4. The other should show temperature versus time for ice melting.
5. Time should be recorded in 0.5-minute (meaning every 30 seconds) intervals.
6. Begin to measure water going from room temperature to a rolling boil, while you add heat to the beaker. The water should be allowed to maintain its boil for three minutes, while you continue measuring temperature.
7. At the same time, measure water with two ice cubes in it going to water with all the ice melted, while you add heat to the beaker (students may also be able to use snow and ice mixed from outside for this portion). The water should continue to be measured for temperature until all the ice has been melted for three minutes.
8. Insert and label two images, showing your apparatus for both experiments in the two boxes on p. 237.
9. Make a line graph showing the data from both experiments with temperature versus time (Figure 6.5, p. 238)

Analysis

1. Write a working definition of *freezing*.
2. Write a working definition of *melting*.
3. Write a working definition of *boiling*.
4. Write a working definition of *condensing*.
5. Does the addition of heat always raise the temperature of water? Give details from your data in your answer.
6. Describe exactly what is happening when the water changes phase.
    a. From liquid to gas:
    b. From solid to liquid:
7. (Enrichment) Give a working definition of *heat*, using the word *molecules* in your answer.
8. (Enrichment) Give a working definition of *temperature*, using the word *molecules* in your answer.

Learning Target

The two main things that happen when heat is added to water are that the temperature of the water increases and a phase change occurs. The two do not happen at the same time, however.

I Learned:

Redo:

Manipulated Variable:

Measured Variable:

Controlled Variable:

INSERT LABELED IMAGE FROM PROCEDURE #6 HERE.

INSERT LABELED IMAGE FROM PROCEDURE #7 HERE.

(*Teacher note:* An engineering design extension for this lab is building an insulator, which can be found on the TryEngineering website: *http://tryengineering.org/lesson-plans*.)

FIGURE 6.5. ______________________________

(*Teacher Note:* Students provide descriptive graph title.)

Reinforcement of Learning

FIGURE 6.6.

MELTING AND BOILING POINT GRAPH OF A PURE SUBSTANCE

(Use completed graph in Figure 6.5 for questions 1–9.)

1. What temperature is the melting point? Explain how you know.
2. What temperature is the boiling point? Explain how you know.
3. What temperature is the freezing point? Explain how you know.
4. What temperature is the condensing point? Explain how you know.
5. What letter represents the range where the liquid is being warmed?
6. What letter represents the range where the solid is being warmed?
7. What letter represents the range where the vapor is being warmed?
8. What letter represents where the substance is boiling?
9. What letter represents where the substance is melting?
10. Why are we measuring temperature in °C during this lab?
11. Sketch what a constant temperature graph would look like.
12. Sketch what an increasing temperature graph would look like.
13. Sketch what a decreasing temperature graph would look like.
14. What is sublimation?
15. (Enrichment) Describe what happens when a water ice cube and solid carbon dioxide are both placed on the lab counter. Then explain why the solid carbon dioxide is called "dry ice."

    (*Teacher note:* Dry ice demonstrations can be added to enhance the analysis section on this lab. Follow appropriate safety precautions, however. Dry ice temperature is extremely cold (about –90°C). Handle dry ice carefully and wear protective gloves when touching it. Store dry ice in an insulated but not completely airtight container. Keep proper air ventilation wherever dry ice is stored. Normal air is 78% nitrogen, 21% oxygen and only 0.035%

carbon dioxide. If the concentration of carbon dioxide in the air rises above 0.5%, carbon dioxide can become dangerous.)

16. (Enrichment) Investigate what the addition of salt does to the temperature versus time graph of phase changes in water. Does salt water have a different result?

17. (Enrichment) What if ice did not float on liquid water? Water is quite unique in that its solid form is less dense than its liquid form. Some scientists have hypothesized that life as we know it could not exist if this unique pattern were not true. Imagine and explain the ramifications for our planet. Use the *ComicBook!* app.

6C. Phasing in Changes Lab

NGSS Alignment

MS-PS1-4. Develop a model that predicts and describes changes in particle motion, temperature, and state of a pure substance when thermal energy is added or removed.

MS-ETS1-1. Define the criteria and constraints of a design problem with sufficient precision to ensure a successful solution, taking into account relevant scientific principles and potential impacts on people and the natural environment that may limit possible solutions.

MS-ETS1-2. Evaluate competing design solutions using a systematic process to determine how well they meet the criteria and constraints of the problem.

6D.

Deciphering a Weather Map Lab

Problem

What do the symbols mean on a weather map?

Prediction

Describe what types of information a weather station collects.

Vocabulary for Glossary

1. *Atmospheric pressure:* Measured by barometer in millibars (mb) and inches of mercury (inHg). Sea Level Average is 1013 mb or 29.92 inHg. On a weather map, "056" refers to 1005.6 mb.
2. *Wind speed:* Measured by anemometer and shown by lines. Full line = 10 knots (kt). Shorter line = 5 knots. Add all lines to get total (FYI: 1 kt = 1.15 mph).
3. *Wind direction:* Measured by wind vane, pointing to the direction from which the wind is blowing (e.g., "from the south").
4. *Temperature:* Measured by thermometer in units of °F for meteorology and °C for routine science class use.
5. *Dew point:* measured by sling psychrometer. The closer the dew point is to the actual temperature, the more humid it feels.
6. *Cloud cover:* Shown by amount of circle, in weather symbol, that is darkened.
7. *Precipitation:* Measured by rain gauge. Symbols show current type and level of rain, snow, sleet, or hail.

Thinking About the Problem

What do we need to know in order to interpret weather maps? These maps report meteorological data collected from several weather stations at a specific point in time. Weather stations can be in many places, including airports, TV and radio broadcasting stations, schools, private homes, and remote areas maintained by the National Oceanic and Atmospheric Administration.

Normally weather maps show an outline of a specific area (local, state, national), the cities where the reporting stations are located, and symbols to show what the weather is like in each city. By including information from a number of different stations, the map will give a good idea of what the weather is across the whole area represented.

Figure 6.7 (p. 244) shows an example of a weather station's symbol and the information given by each part of the symbol. Although this current system may change, weather station symbols in the United States are still expressed with the English Standard units of measurement, rather than the metric system.

Write three main points from the "Thinking About the Problem" reading:

1. Weather station symbols in the United States are still expressed with ...

2.

3.

Analysis

(*Teacher note:* Answers are in parentheses. Students use Figures 6.7–6.13 [pp. 244 –247] to find the answers.)

1. What is the current precipitation in Pueblo, Colorado? (Slight showers)
2. What is the atmospheric pressure in Miami, Florida? (1006.7 mb)
3. What is the wind direction and wind speed in Winnipeg, Manitoba? (20 kts from the NW)
4. What is the temperature in San Antonio, Texas? (65°F)

5. What is the cloud cover in Boston, Massachusetts? (100% cloudy)
6. What is the atmospheric pressure in Phoenix, Arizona? (1005.9 mb)
7. What is the precipitation in Chicago, Illinois? (Thunderstorms)
8. What is the cloud cover in Seattle, Washington? (100% clear)
9. Describe the wind direction and wind speed in Los Angeles, California. (30 kts from the NW)
10. Describe the precipitation in Minneapolis. (Slight showers)
11. Describe the region of the map that appears to be generally cloudy. (Midwest through the Northeast)
12. Describe the region of the map that appears to be generally clear. (Northwest and southern)
13. Describe a region in the map that fits your ideal weather, and explain why. (Answers vary)
14. In Sciencerocks, the weather has changed to the following conditions: atmospheric pressure is 1015 mb; temperature is 54°F; dew point is 40°F; wind is 25 knots from the SE; and cloud cover is 50%. Draw and label a sketch of the current weather symbol for Sciencerocks in the box on p. 244.
15. (Enrichment) Why is it important to be informed about weather conditions? Describe 10 careers that rely on forecasted weather conditions.

Learning Target

Develop an understanding of weather station information to interpret weather maps.

I Learned:

Redo:

Manipulated Variable:

Measured Variable:

Controlled Variable:

LABELED SKETCH FOR ANALYSIS #14

FIGURE 6.7.

WEATHER STATION FOR THE CITY OF "SCIENCEROCKS"

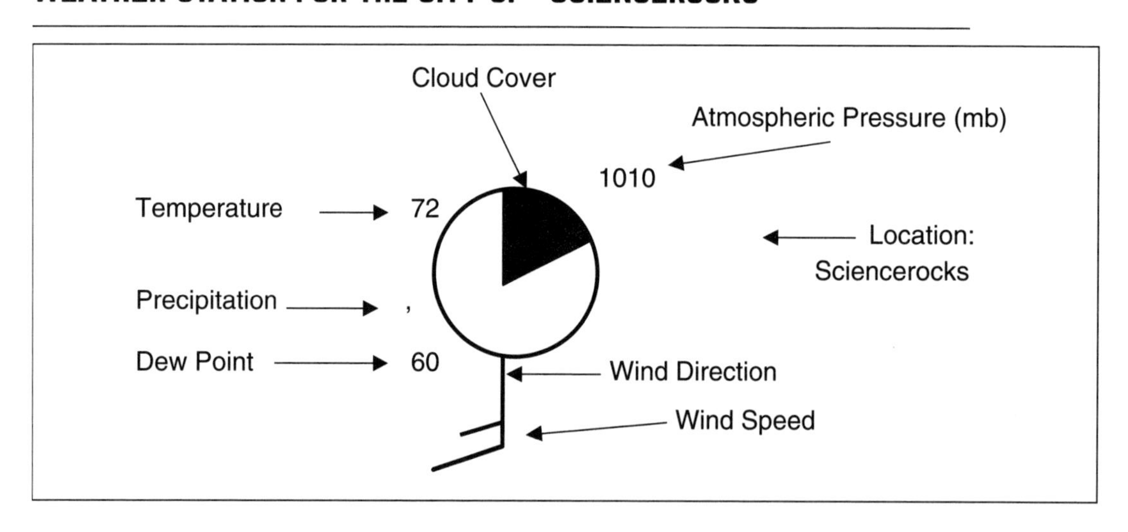

FIGURE 6.8.

WEATHER SYMBOLS FOR WIND SPEEDS

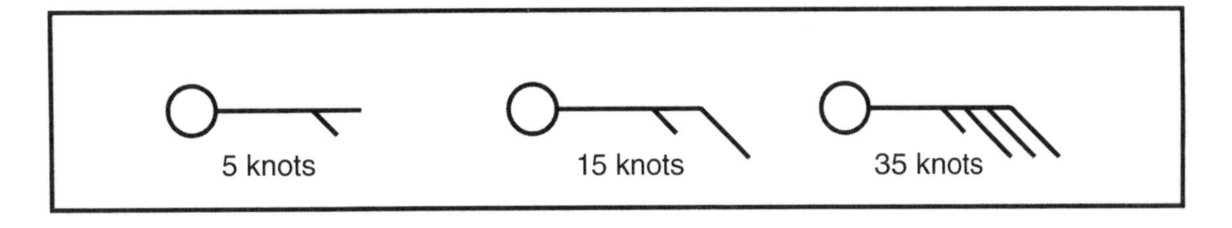

FIGURE 6.9.

WEATHER SYMBOLS FOR WIND DIRECTION

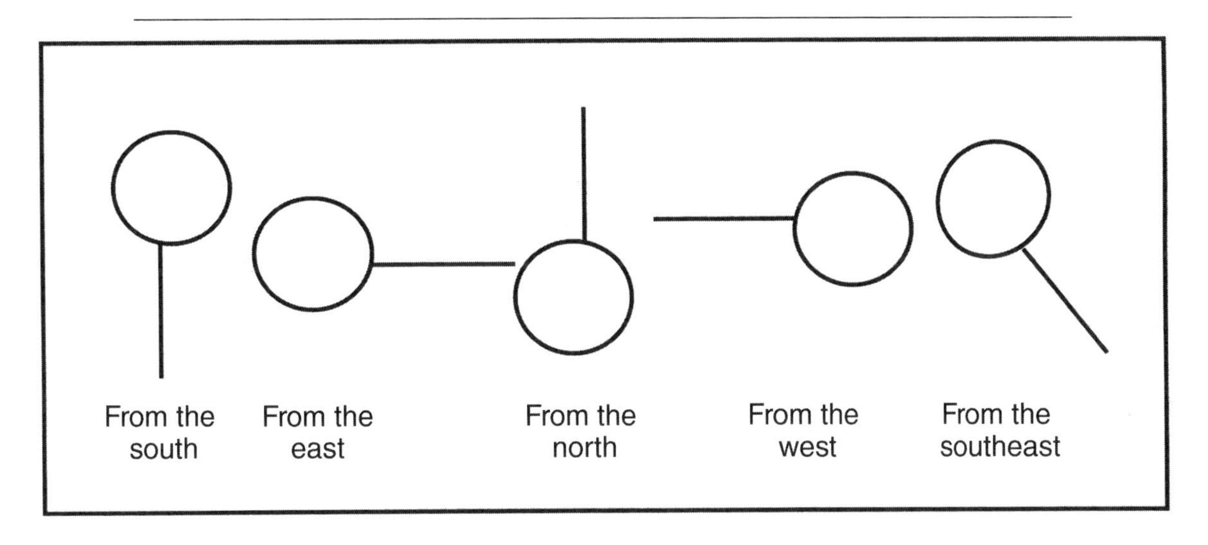

FIGURE 6.10.

WEATHER SYMBOLS FOR CLOUD COVER

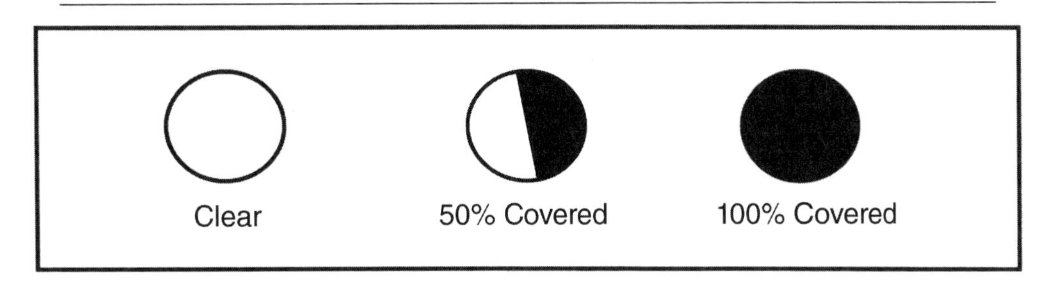

FIGURE 6.11.

WEATHER SYMBOLS FOR PRECIPITATION

SYMBOL	PRECIPITATION	SYMBOL	PRECIPITATION
•	Intermittent rain	,	Intermittent drizzle
• •	Continuous rain	, ,	Continuous drizzle
Δ	Hail	⌈⚡	Thunderstorms
Δ•	Sleet	=	Fog
*	Intermittent snow	$\overset{\circ}{\nabla}$	Slight showers
* *	Continuous snow	$\overset{\circ}{\overline{\nabla}}$	Heavy showers

FIGURE 6.12.

NATIONAL MAP FOR WEATHER STATION SYMBOLS

FIGURE 6.13.

WEATHER STATION SYMBOLS FOR EACH CITY ON THE NATIONAL MAP

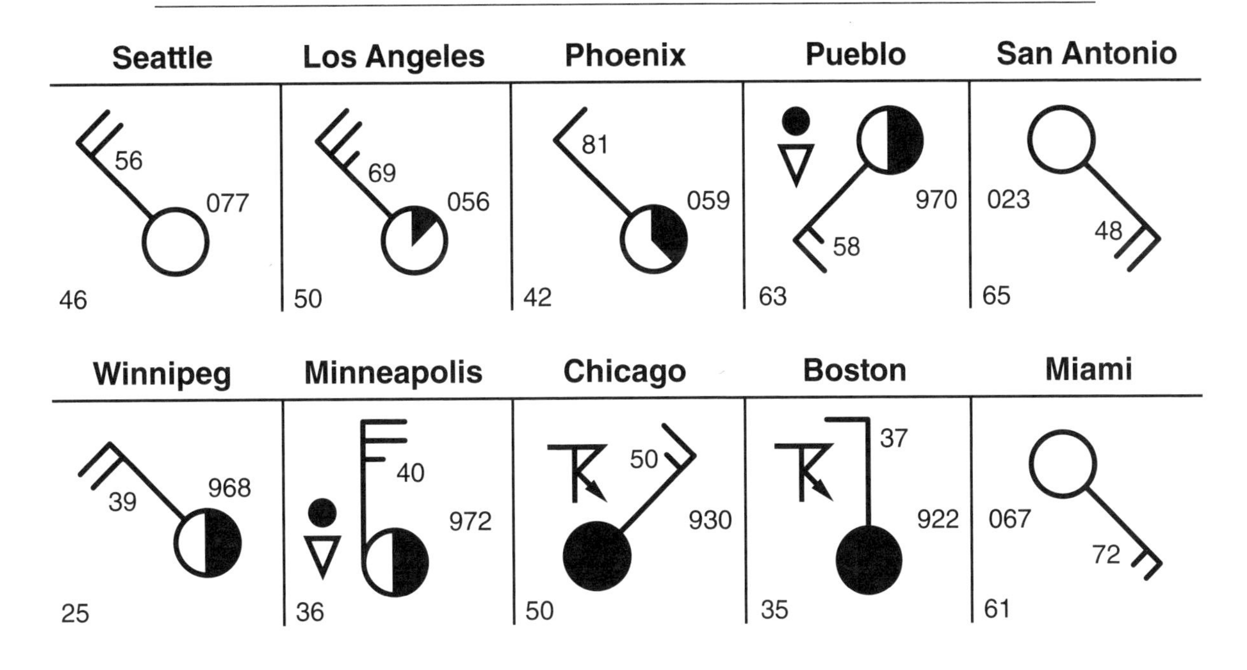

6D. Deciphering a Weather Map Lab

NGSS **Alignment**

MS-ESS2-5. Collect data to provide evidence for how the motions and complex interactions of air masses results in changes in weather conditions.

6E.

Wednesday Weather Watch Reports

Learning Target

Learn the effects that air masses, humidity, air pressure, and wind have on our daily weather.

Requirements

1. Present current weather for Wednesday and a forecast for the week to the class.
2. You will work in a small group for this presentation. It is not a group grade, however. You will earn your own grade, through completion of the requirements on this worksheet.
3. On your assigned day you must show this form, with Part B completed, to your teacher. Bring this form, with the required forecast information filled out using the National Weather Service (NWS) in our local area (from within the past 24 hours.).
4. You will use our school's weather station, to obtain all relevant current information (Part A) while you are in class.

Presentation (Part A)

1. The following information is the current weather data from the weather station in our classroom.
2. The current temperature is: ___
3. The current air pressure is: ___ and it is ___ (Rising? Falling? Steady?)
4. The current humidity is: ___
5. The current wind speed is: ___

6. The current wind direction is from the ___
7. The current precipitation type is: ___ (Look Outside: rain, snow, cloudy, clear skies)

Presentation (Part B)

1. The following information comes from the National Weather Service in ___
2. The forecast for today is: ___
3. The forecast for Thursday is: ___
4. The forecast for Friday is: ___
5. The forecast for Saturday is: ___
6. The forecast for Sunday is: ___

Teacher note: The National Oceanic and Atmospheric Association has a high-resolution modeling system that produces high-quality, colorful weather maps. You may want to investigate the following websites to learn more:

- Educational resources: *http://esrl.noaa.gov/outreach/education.html*
- Weather maps:
    - *www.wired.com/2014/10/hrrr-weather-map*
    - *http://ruc.noaa.gov/hrrr*
    - *http://esrl.noaa.gov*

6E. Wednesday Weather Watch Reports

NGSS Alignment

MS-ESS2-5. Collect data to provide evidence for how the motions and complex interactions of air masses results in changes in weather conditions.

6F.

Lining up in Front Lab

Problem

How do warm and cold fronts influence the weather you see?

Prediction

Give an answer, in one complete sentence, to the problem statement.

(*Teacher note:* Have students share several predictions out loud, so misconceptions can be anticipated and explained.)

Thinking About the Problem

What is a weather front? When a warm air mass meets a cool air mass, two distinct bodies of air are brought in contact, each with its own temperature and relative humidity. Normally, when this "collision" happens, the warm air mass rides up and over the cool air mass, because cool air is denser, and basically shoves the warm air out of its way. A front is, then, the boundary or line between these two air masses.

When a weather front results in warm air getting pushed above cool air, the warm air mass swells as it goes higher and higher. This happens because the air pressure is lower at higher altitudes. As the warm air rises and expands, it begins to cool and the moisture it contains condenses to form clouds and often rain, ice, or snow. Clouds come in different shapes and sizes and this is a factor that meteorologists use to help predict the weather.

Dew is the result of air reaching a certain temperature at which it becomes saturated. Saturation occurs when the air can hold no more water vapor. This vapor will begin to change to liquid, as, for example, often happens in early morning, when moist air condenses on cooler grass, rocks, and trees. If the temperature of the grass (for example) is below freezing, the condensation (*densare* "to thicken" in Latin) is known as frost. The temperature at which the processes of evaporation and condensation are equal is called the dew point.

Write three main points from the "Thinking About the Problem" reading:

1. Dew Point is ...

2.

3.

Procedure

1. Copy Figure 6.15 (p. 253). Separate the three strips, cutting along the lines. Also cut out the "city."
2. Shade in the cold air masses, so they are easier to see.
3. Glue the three strips together by matching up the letters.
4. To make the viewer (as shown in Figure 6.14), fold a piece of notebook paper in half and make two 6 cm cuts that are vertical, as shown below.

FIGURE 6.14.
VIEWER

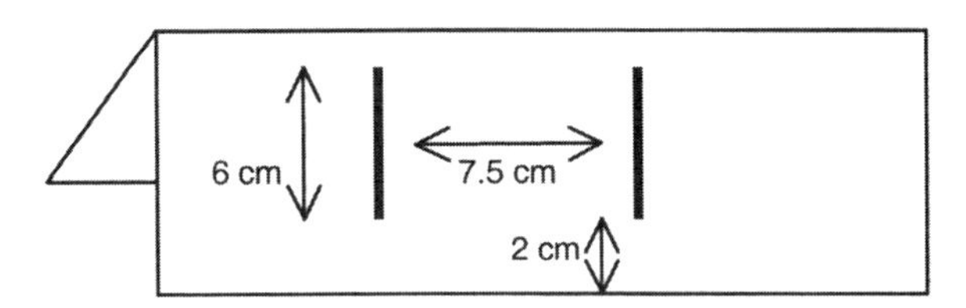

5. Glue/tape/staple the city below the two cuts.
6. Feed the strip through the two cuts so that it passes over the city, starting with Zero Hours.

Analysis

1. Describe what the clouds looked like during the first few hours.
2. What did the clouds look like after 24 hours, when it was raining?
3. Why did the warm air mass rise up over the cold air mass?
4. Describe the types of clouds present as the cold front moved into the city at the end.
5. If you saw wispy clouds followed by lower layered clouds, what type of weather might you expect in the next 24 hours?
6. (Enrichment) Briefly research and report on the weather proverb, "*Mare's tails* and *mackerel* scales *make tall ships carry low sails.*" What does the proverb mean, and exactly what would the cloud conditions be if it was applicable?
7. (Enrichment) Research the "cloud in a jar" demonstrations on the internet. Using the simple materials required, prepare a demonstration in which you generate a cloud in a jar, and ask someone to video it using your iPad. Show the video to the class.

Learning Target

Determine how warm and cold fronts influence the weather that we experience.

I Learned:

Redo:

Manipulated Variable:

Measured Variable:

Controlled Variable:

FIGURE 6.15.

LINING UP IN FRONT STRIP

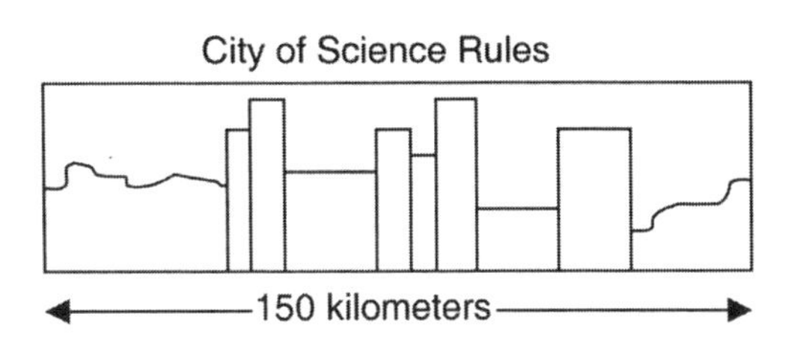

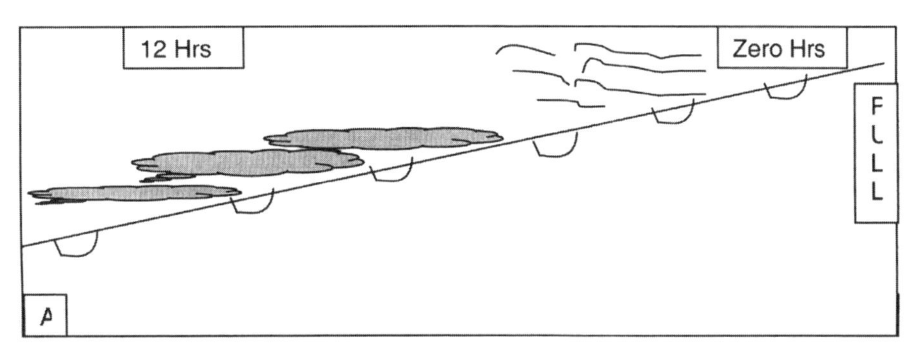

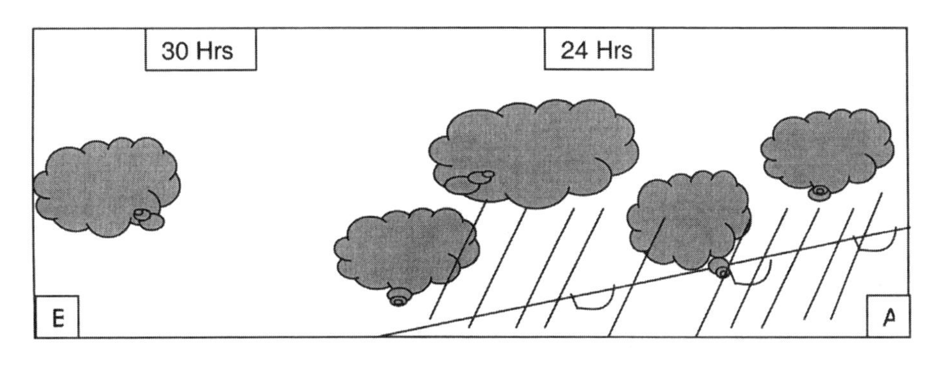

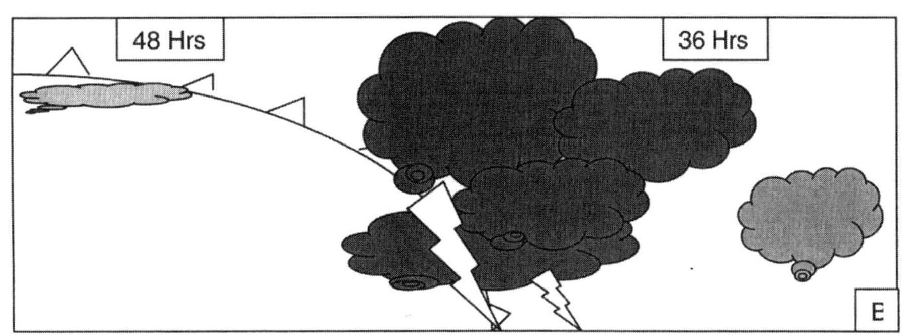

6F. Lining up in Front Lab

NGSS Alignment

MS-ESS2-5. Collect data to provide evidence for how the motions and complex interactions of air masses results in changes in weather conditions.

6G.

Weather Instrument Project

This is an at-home project. It will follow the curriculum for our study of meteorology in Earth science. Students will begin this project by selecting one of the following weather instruments listed below. (Three of the instruments provide less challenge, and therefore result in a slightly lower grade.) This project is to be done individually, without partners.

Instrument Choices

1. Anemometer
2. Barometer
3. Hygrometer
4. Precipitation gauge (The highest grade possible is "B," 85%.)
5. Thermometer
6. Wind vane (The highest grade possible is "B," 85%.)
7. Sling psychrometer
8. Weather stone (The highest grade possible is "C," 75%.) See Figure 6.16.

FIGURE 6.16.

SAMPLE WEATHER STONE

Science Rocks' Weather Stone	
Stone is wet: Rain	Stone is dry: Not raining
Shadow on ground: Sunny	White on top: Snowing
Can't see stone: Foggy	Stone is moving: Windy
Stone is jumping up and down: Earthquake	Stone is gone: Tornado

Student Tasks

1. You will build (not purchase) a particular weather instrument.
2. You will conduct research on that instrument, in order to complete a "12 Facts" report.
3. You will use the instrument you built to record measurements of current weather conditions for five consecutive days.
4. You will display one data table (Data Table 6.4) and one graph of your results.
5. You will share the weather instrument you built, along with a picture you took of it while it was operating.

Required Research

Students may use internet search engines or library resources to first research the following information. Students will use this research to complete the "12 Facts" report.

1. Find what the weather instrument is used to measure (define its purpose).
2. Determine how the weather instrument is engineered (describe the mechanics of how it works).
3. Research how to build your own version of that weather instrument, so you can build it.
4. Research a scientist from history who invented, or is known for first using, this particular weather instrument.
5. On the due date, the project will be presented to classmates on using a gallery tour with both the instrument and the standalone folder opened for display.

Data Collection

After designing and constructing the instrument, you will use it to record the measurements for current weather conditions. Students must take measurements at least three different times per day for five consecutive days. Data must be collected that can be recorded on both a data table and a graph. This data should include verification for accuracy by checking and comparing with actual weather instrument

data (use internet sites, television and radio broadcasts, our classroom weather stations, weather station websites, and newspapers).

The instrument and a standalone folder (all information must fit onto a manila file folder, which is opened for review in Figure 6.17) showing research results will be due. At least one data table of results, at least one graph of results, and at least one picture of your weather instrument, must be included on your folder.

On the gallery tour for project presentation day, students will complete their Weather Instrument "I Learned…" worksheet (Data Table 6.5, p. 259) to show individual accountability, and they will write one encouraging statement on each Weather Instrument Affirmations form (Figure 6.18, p. 260) while they travel around the classroom.

FIGURE 6.17.

STANDALONE FOLDER

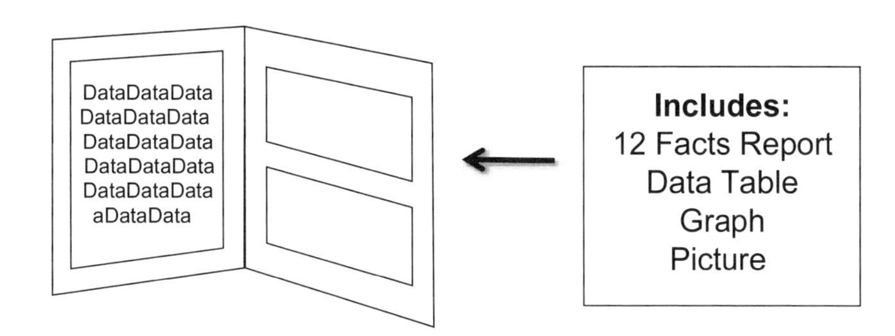

(*Teacher note:* Engineering design lesson plans can be found for hygrometers and anemometers at the TryEngineering website: *http://tryengineering.org/lesson-plans*.)

DATA TABLE 6.4.

SAMPLE DATA TABLE (TO HELP GET YOU STARTED)

DATE	TIME OF DAY	WEATHER INSTRUMENT READING	VERIFIED ACTUAL READINGS	OTHER OBSERVATIONS

12 Facts Report

Complete each of the following sentence starters, using your research, in a document. Display this word document, along with your data table, graph, and picture, in the standalone folder.

1. The weather instrument that I chose to construct is …
2. Its purpose is…
3. The materials used to construct the instrument are …
4. The mechanics of the real-world weather instrument work by …
5. Three facts about the inventor's life story (family, education, birthplace, etc.) include ...
6. The original inventor's weather instrument is different from mine in the following way …
7. The year that this weather instrument was invented is …
8. The invention makes our lives easier, safer, or healthier in the following way …
9. The data that I collected includes all of the following …
10. The graph of my data shows …
11. If I were to do this project over again, I would …
12. If I could modify the construction of my weather instrument, I would …

DATA TABLE 6.5.

WEATHER INSTRUMENT "I LEARNED ..." WORKSHEET

STUDENT NAME	INSTRUMENT	"I LEARNED ..."

FIGURE 6.18.

WEATHER INSTRUMENT AFFIRMATIONS FORM

Weather Instrument
Affirmations

Student Name: ___________________________

Hour: __________

Instrument Chosen: _______________________

6G. Weather Instrument Project

NGSS Alignment

MS-ESS2-5. Collect data to provide evidence for how the motions and complex interactions of air masses results in changes in weather conditions.

MS-ETS1-1. Define the criteria and constraints of a design problem with sufficient precision to ensure a successful solution, taking into account relevant scientific principles and potential impacts on people and the natural environment that may limit possible solutions.

MS-ETS1-2. Evaluate competing design solutions using a systematic process to determine how well they meet the criteria and constraints of the problem.

MS-ETS1-3. Analyze data from tests to determine similarities and differences among several design solutions to identify the best characteristics of each that can be combined into a new solution to better meet the criteria for success.

MS-ETS1-4. Develop a model to generate data for iterative testing and modification of a proposed object, tool, or process such that an optimal design can be achieved.

3-5-ETS1-2. Generate and compare multiple possible solutions to a problem based on how well each is likely to meet the criteria and constraints of the problem.

6H.

Making Your Own Cloud Chart

Directions

1. Begin by doing the Virtual Cloud Lab from PBS, which allows you to try classifying clouds and investigating the role they play in severe storms (*www.pbs.org/wgbh/nova/labs/lab/cloud*)
2. Take your own picture of the clouds in our sky, here in our community. Take a great picture (clear, in focus, spectacular, impressive)! Paste it below, to fill the box on this assignment. It must show some easily recognizable landmark in our community. Be creative with what landmark you will include. Label the picture with the cloud type(s), the date and time that the picture was taken, and the name of the landmark shown in the picture.

INSERT IMAGE OF CLOUDS IN YOUR COMMUNITY HERE.

3. Complete Cloud Chart Data Table 6.6 (p. 264), using research and images from the internet.

4. Draw or sketch (no images) and label your own version of an atmosphere diagram, showing which cloud types can be found at particular altitudes. An incomplete sample is shown (Figure 6.19, p. 265). Draw your own diagram, which includes the nine cloud types shown in the Cloud Chart Data Table 6.6 (p. 264).

DATA TABLE 6.6.

CLOUD CHART

ALTITUDE	CLOUD TYPE	CLOUD IMAGE FROM INTERNET	DESCRIPTION OF ASSOCIATED WEATHER
High-Altitude Clouds	Cirrus		
	Cirrocumulus		
	Cirrostratus		
Mid-Altitude Clouds	Altocumulus		
	Altostratus		
Low-Altitude Clouds	Cumulus		
	Stratus		
	Nimbostratus		
All Levels	Cumulonimbus		

FIGURE 6.19.

DRAWING OF ATMOSPHERE LAYERS DIAGRAM

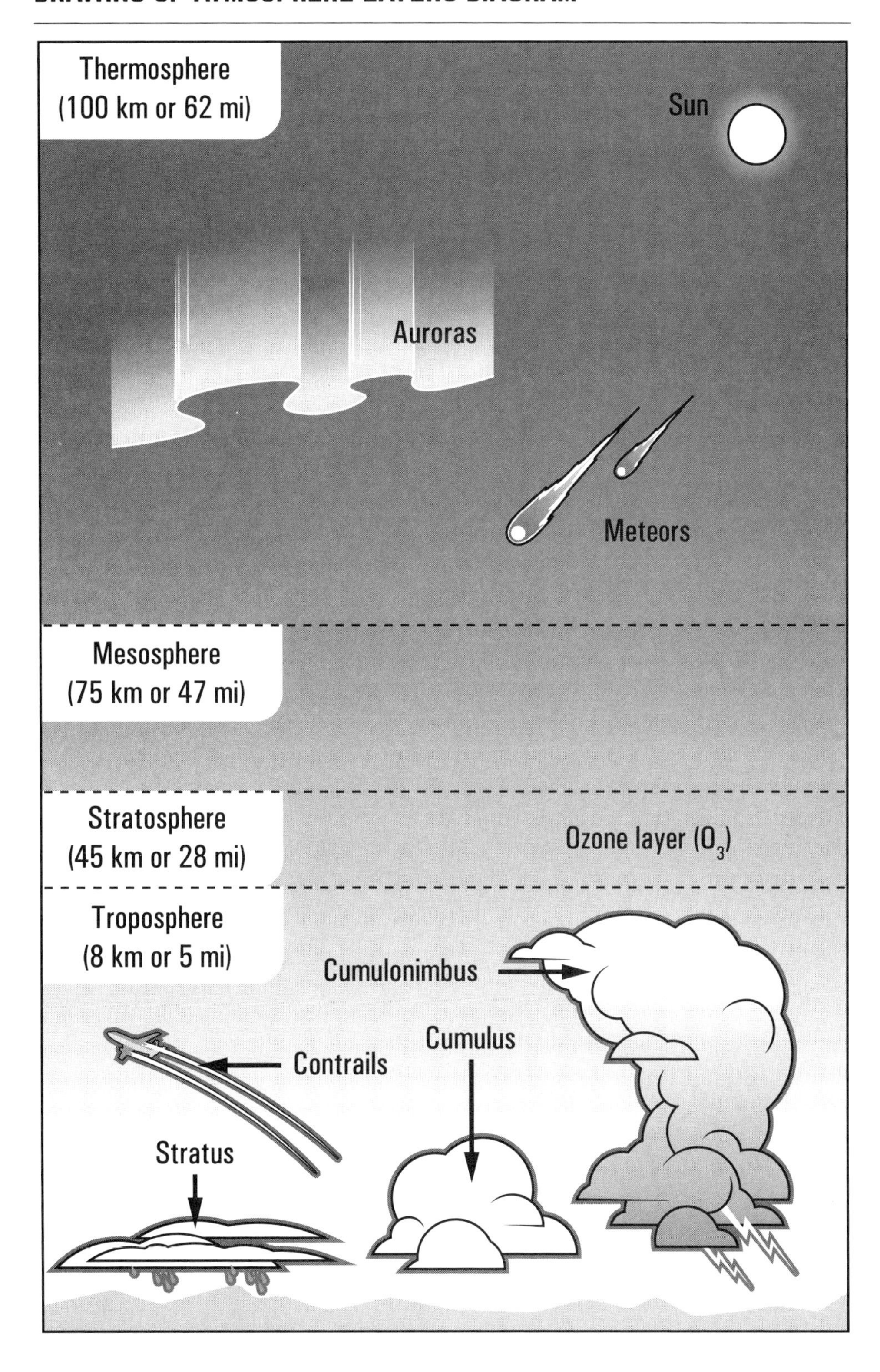

Teacher note: An engineering design tie-in can be found in the TryEngineering website (*http://tryengineering.org/lesson-plans*), which has a great lesson to show students how to build a solar-powered building out of common materials. Additionally, introduce the greenhouse effect now by showing students the two-minute animation from Howard Hughes Medical Institute's BioInteractive site (*http://goo.gl/Vd26U4*).

6H. Making Your Own Cloud Chart

NGSS Alignment

MS-ESS2-5. Collect data to provide evidence for how the motions and complex interactions of air masses results in changes in weather conditions.

MS-ETS1-1. Define the criteria and constraints of a design problem with sufficient precision to ensure a successful solution, taking into account relevant scientific principles and potential impacts on people and the natural environment that may limit possible solutions.

MS-ETS1-2. Evaluate competing design solutions using a systematic process to determine how well they meet the criteria and constraints of the problem.

61.

Weather Proverbs Presentation

Directions

A number of fairly reliable weather proverbs are based upon observations and apply particularly well to the mid-latitudes of the Northern Hemisphere. They are related to the succession of clouds that occur as warm or cold fronts approach, to the drop in air pressure, and to the general movement of weather changes from west to east. Your task will be to add to this list, with your own proverb.

Examples

1. Red sky at night, sailors delight;
   Red sky at morning, sailors take warning.
2. When the grass is dry at morning light,
   Look for rain before the night.
   When the dew is on the grass,
   Rain will never come to pass.
3. When the ditch and pond offend the nose,
   Then look for rain and stormy blows.
4. Rain before seven, shine before eleven.
5. When a cow's tail faces west,
   Then the weather will be best.
6. A rainbow in the morning is the shepherd's warning.
   A rainbow at night is the shepherd's delight.
7. Trace in the sky the painter's brush;
   The winds around you will soon rush.
8. If you see long contrails in our atmosphere,
   Hide in fear, the sky won't be clear. (This one is by Awil M., a former student of the author.)

Weather Proverbs Presentation Requirements

1. This presentation will be accomplished using *Explain Everything* app or *Notability* app, to be AirServer shown to class on presentation date.
2. Must be a poem (easy to remember, rhyming, and so on.)
3. Must have at least four lines.
4. Must be colorful, easy to read, and grammatically correct.
5. Must use at least one colorful image.
6. Must be an accurate proverb about our weather.
7. Students may work with one partner. Students must creatively develop their proverb.

61. Weather Proverbs Presentation

NGSS Alignment

MS-ESS2-5. Collect data to provide evidence for how the motions and complex interactions of air masses results in changes in weather conditions.

6J. Sweating About Science Lab

Problem

Is there a difference between indoor humidity and outdoor humidity?

Prediction

Give a working definition of *relative humidity*.

(*Teacher note:* Have students share several predictions out loud, so misconceptions can be anticipated and explained.)

Thinking About the Problem

Why do we perspire more when it is humid? This is actually a misconception that many people share. The truth has to do with evaporation. In dry air, it is easier for water to evaporate and enter the air as water vapor. On very humid days, the air is already holding as much water vapor as it can, so perspiration is unable to evaporate and sits on our skin.

Hot, dry air is normally more comfortable than warm, humid air. Similarly, cold, dry air is often more comfortable than cool, damp air. While there is always at least a small amount of moisture in the air we breathe, the amount makes a big difference on our comfort level. Why? Temperature plays the primary role in determining the amount of water vapor present in the air at any given time. Warm air can hold more water vapor than cool air. Every temperature has its limit where it can hold no more water vapor. Saturated air has a relative humidity of 100% and clouds or fog begin to form. When relative humidity is high, our perspiration cannot evaporate quickly, and we become uncomfortable.

Relative humidity is a measure of how much water vapor is actually in the air compared to the amount the air could possibly hold. In desert areas, relative humidity is low, but it is high in jungles. Relative humidity can be measured with two different instruments. One is called a hygrometer (*hugros* "wet, moist" in Greek); the other is called a sling psychrometer (*psukhros* "cold" in Greek). Since the sling psychrometer is easily portable, that is the one we will use in this lab.

Write three main points from the "Thinking About the Problem" reading:

1. A sling psychrometer measures ...

2.

3.

Procedure

1. Read the "dry bulb temperature" on the metal thermometer in °F. Convert temperatures using the converter from the National Oceanic and Atmospheric Association (*http://goo.gl/Wa8X4U*, or scan the QR code below), then record the dry bulb temperature in °C. Draw your lab materials in box on p. 271.

QR CODE FOR THE TEMPERATURE CONVERTER

2. Dip the wet bulb in your beaker of water, and then dry off the dry bulb. Carefully swing the sling psychrometer around for two minutes (use the clock app on your tablet to time it accurately), recording the lowest temperature reached by the thermometer. Convert this "wet bulb temperature" from °F to °C and record.
3. Subtract the wet bulb temperature from the dry bulb temperature, and record.
4. From your calculation in step 3, determine the relative humidity using Data Table 6.7.

5. Use this same method to determine the relative humidity in four more locations around the school.

INSERT LABELED IMAGE OF LAB MATERIALS HERE.

Analysis

1. What did you notice about the wet bulb temperature while you were swinging the sling psychrometer?
2. Give a good explanation for what you observed in question 1 above.
3. Was there a difference between the indoor and outdoor relative humidity? Explain.
4. (Enrichment) Briefly research the relative humidity of both Colorado Springs, Colorado, and Phoenix, Arizona, during the last 20 years. How does construction of residential areas, landscaping around businesses, and city development affect or influence relative humidity?
5. (Enrichment) Some people's hair curls when the relative humidity is high. Think of a way to use this fact to measure relative humidity. Include a labeled sketch of the device you would use with your explanation.

Learning Target

Use weather instruments to determine the difference between indoor and outdoor relative humidity.

I Learned:

Redo:

Manipulated Variable:

Measured Variable:

Controlled Variable:

DATA TABLE 6.7.

RELATIVE HUMIDITY DATA

LOCATION	EXACT TIME OF DAY	DRY BULB TEMP (°C)	WET BULB TEMP (°C)	DIFFERENCE (°C)	RELATIVE HUMIDITY (%)
Our classroom					

DATA TABLE 6 8.

PERCENTAGE OF RELATIVE HUMIDITY

DIFFERENCE (°C)	DRY BULB TEMP (°C)																														
	5	6	7	8	9	10	11	12	13	14	15	16	17	18	19	20	21	22	23	24	25	26	27	28	29	30	31	32	33	34	35
1	86	86	87	87	88	88	89	89	90	90	90	90	90	91	91	91	92	92	92	92	92	92	92	93	93	93	93	93	93	93	94
2	72	73	74	75	76	77	78	78	79	79	80	81	81	82	82	83	83	83	84	84	84	85	85	85	86	86	86	86	87	87	87
3	58	60	62	63	64	66	67	68	69	70	71	71	72	73	74	74	75	76	76	77	77	78	78	78	79	79	80	80	80	81	81
4	45	48	50	51	53	55	56	58	59	60	61	63	64	65	65	66	67	68	69	69	70	71	71	72	72	73	73	74	74	75	75
5	33	35	38	40	42	44	46	48	50	51	53	54	55	57	58	59	60	61	62	62	63	64	65	65	66	67	67	68	68	69	69
6	20	24	26	29	32	34	36	39	41	42	44	46	47	49	50	51	53	54	55	56	57	58	58	59	60	61	61	62	63	63	64
7	7	11	15	19	22	24	27	29	32	34	36	38	40	41	43	44	46	47	48	49	50	51	52	53	54	55	56	57	57	58	59
8				8	12	15	18	21	23	26	27	30	32	34	36	37	39	40	42	43	44	46	47	48	49	50	51	51	52	53	54
9						6	9	12	15	18	20	23	25	27	29	31	32	34	36	37	39	40	41	42	43	44	45	46	47	48	49
10									7	10	13	15	18	20	22	24	26	28	30	31	33	34	36	37	38	39	40	41	42	43	44
11											6	8	11	14	16	18	20	22	24	26	28	29	31	32	33	35	36	37	38	39	40
12														7	10	12	14	17	19	20	22	24	26	27	28	30	31	32	33	35	36

6J. Sweating About Science Lab

NGSS **Alignment**

MS-ESS2-5. Collect data to provide evidence for how the motions and complex interactions of air masses results in changes in weather conditions.

7

HUMAN IMPACTS ON EARTH SYSTEMS

7A. pHiguring out Acids and Bases Lab

Problem

How is the use of pH indicators similar to that of radioactive isotopes?

Prediction

Give a working definition of pH.

(*Teacher note:* Have students share several predictions out loud, so misconceptions can be anticipated and explained. Investigate learning opportunities available through the internet, such as *Citizen Science*, *eCybermission* community connections, *Kidsphere* posted questions, and *Project Circle* electronic discussion groups.)

Thinking About the Problem

Have you ever had a blood test? Blood tests tell much about individuals. They can identify certain diseases and show the level of sugar or alcohol in the body. Athletes are given blood tests to be sure they are not using steroids. Many uses of blood tests are possible because of the work of Rosalyn Sussman Yalow.

Yalow was born in the Bronx, New York, in 1921. She worked for 22 years developing a way to use radioactive elements to detect certain substances in body tissues. The procedure is called radioimmunoassay (RIA). In 1977, she won the Nobel Prize in Physiology or Medicine for her work. In RIA, radioisotopes (radioactive elements with a different numbers of neutrons) are used indicators to identify various chemical reactions within the body.

Other indicators are used to identify acids and bases in the environment. The term *pH* is used to describe how acidic or basic a substance is. The term refers to the percent of Hydrogen ions (H_3O^+) in a substance.

The values for pH go from 0 to 14. Pure (distilled) water is neutral (neither acid nor base), and is assigned the pH value of 7. Acidic substances have pH values less than 7. Basic substances have pH values greater than 7.

Indicators can be used to identify pH by changing color in the presence of an acid or a base. For example, pink litmus paper turns blue in a base, but remains pink in an acid. Blue litmus paper turns pink in an acid, but does not change color in a base. Other indicators are used to identify the exact pH value of substances. Many environmental chemicals can be used to indicate the presence of an acid or base. Such chemicals include red cabbage juice, which is a deep purple-blue color. In the presence of an acid, it turns red or pink. In the presence of a base, the cabbage juice turns light blue or green-blue. The color change of some indicators identifies the exact pH of a substance.

Write three main points from the "Thinking About the Problem" reading:

1. Indicators are ...

2.

3.

Procedure

1. Wash and carefully dry your hands.
2. Cut both of your pH papers so that you have seven equal pieces. Set them in a dry spot on your lab table.
3. Label seven paper cups with each of the following: water, laundry detergent, window cleaner, carbonated soda, vinegar, all-purpose cleaner, and lemon juice.
4. Using a separate eyedropper for each, place five drops of the appropriate liquid into each cup.
5. Touch a small piece of one pH paper piece to one of the liquids. Record your observations (Data Table 7.1, p. 283).
6. Using a clean eyedropper, add 2–3 drops of red cabbage juice to each cup. Record your observations. Draw your materials in box on p. 282.
7. Turn on your faucet and rinse all liquids from paper cups. Throw away all paper products.

Analysis

1. How is the use of pH indicators similar to that of radioactive isotopes?
2. Which substances, if any, did you identify as neutral?
3. Which substances, if any, did you identify as acids?
4. Which substances, if any, did you identify as bases?
5. Why was it necessary to wash your hands before beginning the experiment?
6. (Enrichment) Steroids and other drugs can be detected in the blood through the use of indicators. Why might such tests be given to athletes? Explain.
7. (Enrichment) Radioisotopes can be detected using a Geiger counter. How might a Geiger counter and radioactive isotopes be used to detect a water main leak?
8. (Enrichment) Compare and contrast the pH values of rain and snow in our community during the course of this school year.
9. (Enrichment) Prepare a public service announcement, using iMovie, on what middle school students should do to improve their environment, and why they should care about its level of health in the first place.

Learning Target

Describe some properties of acids and bases, including their changes with litmus and other indicators.

I Learned:

Redo:

Manipulated Variable:

Measured Variable:

Controlled Variable:

INSERT LABELED IMAGE OF LAB MATERIALS SETUP HERE.

DATA TABLE 7.1. ______________________________

(*Teacher Note:* Student provides descriptive title for this data table.)

SUBSTANCE	GENERAL OBSERVATION	REACTION TO PH PAPER	REACTION TO CABBAGE	ACID OR BASE?
Insert image of pH Paper Color Key in this row.				

7A. pHiguring out Acids and Bases Lab

NGSS Alignment

MS-PS1-1. Develop models to describe the atomic composition of simple molecules and extended structures.

5-ESS3-1. Obtain and combine information about ways individual communities use science ideas to protect the Earth's resources and environment.

7B.

Acid Rain Background Reading

What is acid rain? Acid rain, or acid precipitation, refers to any precipitation having a pH value less than that of normal rainwater. Carbon dioxide (CO_2) in the atmosphere makes normal rain slightly acidic. This is because carbon dioxide and water combine to form carbonic acid, commonly known as carbonated water. Its pH generally ranges from 5.0 to 5.6. Acid rain falling in the northeastern United States is often in the pH range of 4.0 to 4.6. Because "acid rain" includes snow, sleet, hail, dew, frost, and fog, it may also be referred to as acid precipitation.

What is the pH scale? The pH scale, which ranges from 0 to 14, is used to measure whether a substance is acidic or basic. Pure water is neither acidic nor basic. It has a pH of 7, which is neutral. Values on the pH scale below 7 are acidic, and those above 7 are basic. Substances having a pH of 1 (battery acid, for example) are very acidic; those having a pH of 13 (such as lye) are very basic.

The pH scale is logarithmic, with a tenfold difference between each number. If the pH drops from 7 to 6, the acidity is 10 times greater; if it drops from 7 to 5, it is one hundred times greater; from 7 to 4, one thousand times more acidic, and so on. For example, vinegar at pH 3 is 10,000 times more acidic than distilled (neutral) water.

What causes acid rain? The burning of fossil fuels, particularly coal or oil, which contain sulfur, contributes to the acidity of rain. During burning, the sulfur in these fuels combines with oxygen from the atmosphere, forming sulfur dioxide (SO_2); burning also produces nitrogen oxides. The sulfur dioxide and nitrogen oxides are carried up the smokestack by the hot exhaust gases. They may remain mixed in the air for several days. The longer they remain in the air, the greater the likelihood that they will form solutions of sulfuric acid (H_2SO_4), and nitric acid (HNO_3). These acids may be dissolved in droplets of water and carried by winds for many miles.

The largest source of sulfur dioxide and nitrogen oxides are coal-fired electric power generating stations. Industrial processes such as the smelting of sulfide minerals also contribute to the problem. Exhaust gases from cars and trucks also release large amounts of nitrogen oxides into the air, causing serious air pollution in high-traffic areas.

Is acid rain a threat to aquatic ecosystems? In most regions of the United States, lakes and streams can tolerate some extra acidity with no adverse environmental consequences. This is true because most soils are basic in nature, which neutralizes the additional acidity.

However, some parts of the United States and eastern Canada are particularly sensitive to acidity. These regions have thin soils covering bedrock that is low in basic material content. Lakes and streams in these areas have little or no capacity to neutralize acid rain. Parts of the Rocky Mountains and the north-central, southeastern, and northwestern United States are among those susceptible to the effects of acid rain.

Many species of fish, such as rainbow trout, brown trout, smallmouth bass, and minnows, are unable to survive below pH 5. Their eggs and larvae are especially sensitive. As the water becomes more acidic, fewer eggs hatch and fish may not grow to maturity.

Many species of amphibians (frogs, toads, and salamanders) breed in temporary pools that are formed by spring rains and melted snow. These pools may become very acidic. The acidity in these temporary pools may cause deformities and death in the eggs and developing embryos.

What other effects does acid precipitation have? Pollutants that produce acid rain can be carried long distances by the wind. Industrial activity does not have to occur in an acid rain-sensitive area in order for that area to be affected. Waters in Canada are being damaged by acid rain produced by power plants in the United States. Trees at high elevations are particularly susceptible to acid fog. Acid rain and acid snow are dissolving the limestone exteriors of buildings in areas far removed from major industrial activity. Statues and artwork are being eaten away by acid precipitation.

What can you do to help reduce acid rain? We can all help by using fossil fuels more wisely. We can use carpools and mass transit; properly maintain our cars and trucks and their pollution-control devices; be more efficient in heating our homes; and conserve electricity by turning off lights and electrical appliances when they are not in use.

7B. Acid Rain Background Reading

NGSS Alignment

MS-PS1-1. Develop models to describe the atomic composition of simple molecules and extended structures.

MS-ESS3-3. Apply scientific principles to design a method for monitoring and minimizing a human impact on the environment.

7C.

Researching Scientists Project

Option 1: Research a Nontraditional Scientist Assignment

Problem

What contributions have men and women of all races and cultures, including Minnesota Native American tribes and communities, made in scientific inquiry and engineering throughout history?

Thinking About the Problem

The history of underrepresented populations—minorities and women—in science is filled with stories of barriers and roadblocks. It is important to recognize these early pioneers, as well as all of the current trailblazers, because knowing the past helps us to improve the future.

There have been many courageous and persistent minorities and women who have worked to build a foundation for current and future discoveries in science and engineering. Historically, the work and research of minorities and women was used, but their discoveries often went unrecognized.

Historically, many women worked for men in science fields, with the men often getting undeserved recognition for what the women had discovered and designed. In the past, high school girls simply were not allowed to take science courses. Many girls were often discouraged from taking any interest in science because traditionally they had no place or future in it.

There are now many more women and minorities represented in scientific fields and careers. Scientific and technological discoveries are improving as a result. However, women and minorities still have a long way to go to achieve recognition in all phases of scientific and engineering education and careers. For example, engineering, physics, and chemistry departments still lack sufficient women and minority faculty members and role models.

Procedure

1. You may work alone or with one partner. If you decide to choose a partner, you will both get the same grade on this project. Please choose your partner wisely.
2. Homework for Tonight: Conduct a Safari search to select a scientist or engineer. Scientists and engineers will be assigned on a first-come-first-served basis. Select one or two backup names, in case you don't get your first choice. Your tasks involve researching the person, finding useful information, and working on creating and improving your *Keynote* presentation.
3. Grading will follow certain guidelines based on your selection.
   - For a maximum of A- (90%): Select a white, male scientist or engineer.
   - For a maximum of A (due to being more challenging): Select a scientist or engineer from an underrepresented population (nonwhite, female).
   - For one extra credit point: Select a Native American scientist or engineer (must be from Minnesota).
4. Your task is to create a five-slide, one-minute *Keynote* presentation on your iPad about the scientist/engineer you have selected. Your *Keynote* presentation should be shared with the class, and must include the following information:
   A. Slide 1: Scientist or engineer's name, date of birth and death, where he or she lived, and one image.
   B. Slide 2: Reason for why you selected this particular scientist or engineer, including one similarity you share with this scientist or engineer.
   C. Slide 3: Two details or facts about his or her accomplishments, events, discoveries, or inventions.
   D. Slide 4: An image or diagram showing his/her main accomplishment, discovery, or invention.
   E. Slide 5: A detailed listing of all references used (web pages, articles, books, and so on).289

Option 2: Research a Scientist Who Looks Like Me Assignment

Directions

Conduct an internet search to select a scientist or engineer who looks like you. Your tasks involve researching the person, finding useful information, and creating a seven-slide, one-minute *Keynote* presentation on your iPad about the scientist or engineer you have selected. Your *Keynote* presentation should be shared with the class, and must include the following information:

A. Slide 1: The name of your scientist or engineer, with one clear image of him or her, and one clear image of you. (The example that the author uses in role modeling this assignment is Dame Kathleen Yardley Lonsdale.)

B. Slide 2: Use the heading, "This scientist or engineer looks like me and shares my heritage. We are both …" with three specific bullet point details. The author's three examples are that both are Irish in heritage, both are female, and both wear glasses.

C. Slide 3: Use the heading, "This scientist or engineer's background, discoveries, and claims to fame include …" with three or more specific bullet point details. The author's example includes the following three details: Dame Lonsdale made many contributions to the field of x-ray crystallography (allows us to study the structure of molecules, by using x-rays to show what they look like), especially involving the organic molecule of Benzene. She was born in 1903 in County Kildare, Ireland. She was married with three children. She was a Professor of Chemistry at University College in London. She was Head of the Department. She was the first woman inducted into the Royal Society Fellowship. She cared deeply about the social responsibility of science. She graduated from Bedford College (degree in physics) in 1922, where she was on the rowing team.

D. Slide 4: Use the heading, "This scientist or engineer's personality trait that I especially admire is …" giving one specific detail and include another image of the scientist or engineer. The author's example is that Dame Lonsdale generously gave assistance to coworkers and encouraged young people to enter the field of science. She is often quoted

as saying that she would "never refuse an opportunity to speak in schools."

E. Slide 5: Use the heading, "This scientist or engineer is similar to me in that ..." giving one specific detail. The author's example includes that Dame Lonsdale cared both about her job and her family. She also cares about exercising and staying in shape.

F. Slide 6: Use the heading, "This scientist or engineer is different from me in that ..." giving one specific detail. The author's example states that Dame Lonsdale believed women should be in the home to raise their children. Although I respect that, I did not make that choice.

G. Slide 7: Use the heading, "A favorite quote from this scientist or engineer is ..." giving one specific detail. The author's example quotes Dame Lonsdale as saying, "Any country that wants to make full use of all its potential scientists and technologists could do so ... but it must not expect to get the women quite so simply as it gets the men."

7C. Researching Scientists Project

NGSS Alignment

5-ESS3-1. Obtain and combine information about ways individual communities use science ideas to protect the Earth's resources and environment.

7D.

Science Article Reviews

Learning Target

Discover new items of interest, about which scientists are learning.

Directions

Twice each quarter, students will locate and read a quality electronic scientific article from a reputable online source. Look for research that uses longitudinal large data sets, and are peer-reviewed (online sites with an ".edu" address will likely satisfy these requirements). You will then complete a science article report. The purpose of this activity is to get you to read, write, and think about current science topics, research, issues, and discoveries that interest you. The articles you choose can be about any scientific topic (biology, physics, chemistry, geology, astronomy, genetics, meteorology, chemistry, technological applications of scientific research, and so on).

Format for Electronic Science Article Reports

1. Write the title of the article.
2. Source: Exactly where did you find the article? (Which reliable source did you use?)
3. Date of Publication: Article must have been published within accepted dates.
4. For each article you must demonstrate that you read and thought about the article by making note of your questions or comments in the margins, underlining main points and looking up words you don't know. The article, with your notes on it, and this form should be combined in one assignment, using the *Notability* app.
5. After you have finished reading the article, answer the following question in complete sentences: What was the main science-related problem or science-related observation that initially caused this to be a newsworthy article? (Be prepared to share this fact with the class.)

6. In a minimum of eight complete sentences, please summarize the main idea of the article. Do not copy from the article. Put these sentences in your own words. Each sentence must present a completely new idea from the article. If your article is not long enough to accomplish this, then select a better article.
7. Write three detailed, and different, "I learned …" statements about the article.
8. After reading the article, write one question about which you are still curious.
9. From the information in the article, write one prediction you could make about the future. Use one example from the article to support your prediction.

7D. Science Article Reviews

NGSS Alignment

MS-ESS1. Earth's Place in the Universe

MS-PS1. Matter and its Interactions

MS-LS1. From Molecules to Organisms: Structures and Processes

MS-ETS1. Engineering Design

7E.

Oatmeal Raisin Cookie Mining Lab

Problem

How does mining affect both the economy and the environment?

Prediction

Give a working definition of *renewable energy*.

(*Teacher note:* Have students share several predictions out loud, so that misconceptions can be anticipated and explained.)

Thinking About the Problem

There are two main types of energy sources available to us on Earth, renewable and nonrenewable. Renewable energy sources are based on natural cycles that can be replenished during a person's lifetime. Examples of these resources include trees and crops that can be replanted, solar energy, wind energy, hydroelectric power, and geothermal energy.

Nonrenewable energy sources are based on limited reserves that were generated several million years ago by unique conditions. Such reserves will eventually run out as the available deposits are consumed. Fossil fuels, which include coal, oil, and natural gas, are the most common examples of nonrenewable resources. In this lab, we will focus on the coal extraction process, although hydraulic fracturing (as in North Dakota, Oklahoma, Ohio, and Pennsylvania) and the associated frack sand mining (in Wisconsin and Minnesota) also generate a similar need for reclamation.

Coal is made up of the cemented and fossilized remains of ancient plants. For millions of years, buried plants were subjected to heat and pressure to become coal. In our history, coal has been used to power steamships and railroad engines, heat homes and generate heat in steel production. Today the primary use for coal is in generating electrical power.

Coal mining disturbs large areas of land, leading to water quality problems and ecosystem impacts. Burning coal produces large amounts of air pollutants that have been linked to mercury pollution, smog, and global climate change.

Materials

- Oatmeal raisin cookies (donated by students' families)
- Petri dishes
- Spoons
- Toothpicks

Procedure

1. The goal is to use careful techniques to remove all raisins from the cookie, while generating as little damage to the cookie as possible.
2. Use only the toothpicks and spoons, not your fingers, for the mining process.
3. All mining wastes (crumbs) must be counted and captured on the petri dish.
4. Take "Before" and "After" pictures of your cookie during this lab and insert them in the blank boxes on p. 296.

DATA TABLE 7.2.

INCOME VS. COST FOR OATMEAL RAISIN COOKIE MINING

Insert the total for each numbered criteria below in the right-hand column.

PROFIT CALCULATION	
1. Number of Raisins Successfully Mined	
2. Money Earned per Raisin	
3. Total Income From Raisins	
4. Number of Crumbs Generated	
5. Cost per Crumb for Reclamation of Land	
6. Total Cost for Crumbs	
7. Total Profit Earned (Total Income #3 Minus Total Cost #6)	

INSERT LABELED IMAGE OF "BEFORE" MINING HERE.

INSERT LABELED IMAGE OF "AFTER" MINING HERE.

Analysis

1. Define *renewable energy* and give an example.
2. Define *nonrenewable energy* and give an example.
3. Describe why you were or were not able to remove all the raisins from the cookie.
4. Compare this activity with the real impact when mining coal. What happens to plants and animals? Consider the income vs. cost (Data Table 7.2, p. 295).
5. (Enrichment) Conduct research to name three states that have coal deposits, and briefly describe the effect that coal mining has on their economy and their environment.
6. (Enrichment) Compare and contrast frac sand mining of oil (North Dakota) and natural gas (Pennsylvania) with the process of coal mining.

Learning Target

Demonstrate the process of mining and lean about the importance of land reclamation.

I Learned:

Redo:

Manipulated Variable:

Measured Variable:

Controlled Variable:

7E. Oatmeal Raisin Cookie Mining Lab

NGSS Alignment

MS-ESS3-3. Apply scientific principles to design a method for monitoring and minimizing a human impact on the environment.

MS-ESS3-4. Construct an argument supported by evidence for how increases in human population and per-capita consumption of natural resources impact Earth's systems.

MS-ESS2-1. Develop a model to describe the cycling of Earth's materials and the flow of energy that drives this process.

4-ESS3-1. Obtain and combine information to describe that energy and fuels are derived from natural resources and their uses affect the environment.

4-ESS3-2. Generate and compare multiple solutions to reduce the impacts of natural Earth processes on humans.

7F.

The Poetry of Earth Science Project

Your task is to create four poems. The poems must be completely designed and written by you, about any of the Earth Science topics we have covered this year.

Why should you do a good job on this assignment? Writing poetry requires that you become a careful observer; all scientists must possess this skill. Writing poetry helps develop your ease with imaginative language, a precursor to the abstract thinking necessary for success in science. The combination of concept-learning and writing poetry helps you function as a problem-solver, rather than just an information receiver. And, perhaps most importantly, you will learn science better when you are required to compare and contrast, summarize, describe, and interpret—all are important facets of poetic writing.

Each of your four poems must have a title. You will work on your own to complete this assignment. You must be willing to share any two of your poems with the class. You must choose from any of the following poetry patterns:

1. Tanka

Tanka is a type of Japanese poetry that contains 5 lines and 31 syllables arranged in a 5-7-5-7-7 pattern.

Example:

I like all weather
Let's go out and enjoy it
Forever changing
Hide from it on occasion
Nature's own video game.

2. "I used to be ... but now I am ..."

Use three to five sentence starters as a pattern to describe scientific concepts.

Example:

I used to be granite,

But now I am gneiss.

I used to be slate,

But now I am shale.

I used to be just sand,

But now I am sandstone.

3. Diamante

A diamante is a seven-line poem that compares opposites using specific parts of speech. The diamond shape of the finished product gives this poem its name.

Line 1: nouns for the subject

Line 2: two adjectives describing the subject

Line 3: three action terms

Line 4: four nouns, two about the subject, two about its antonym

Line 5: three action terms describing the antonym

Line 6: two adjectives describing the antonym

Line 7: the antonym

Example:

Sun

Hot and radiating

Powerful, energetic, and strong

Gases, gravity Rock, and ice

Revolving, rotating, together for always

Colder and reflecting

Planets

4. Alphabet Pyramid

These are five-line cumulative poems that contain specific parts of speech that begin with the same letters.

Line 1: the letter

Line 2: a noun

Line 3: add an adjective

Line 4: add a verb

Line 5: add an adverb

Example:

C

Contrails

Continuous contrails

Continuous contrails circulate

Continuous contrails circulate convincingly

5. Cinquain

A cinquain is a five-line descriptive poem that contains about 22 syllables.

Line 1: the subject

Line 2: four syllables describing the subject

Line 3: six syllables showing action

Line 4: eight syllables expressing a feeling or observation about the subject

Line 5: two syllables renaming the subject

Example:

Earth

Rocky, water-laden planet

Cycling, changing, and warm

Amazing, fragile, and vital

Our home

7F. The Poetry of Earth Science Project

NGSS Alignment
MS-ESS1. Earth's Place in the Universe

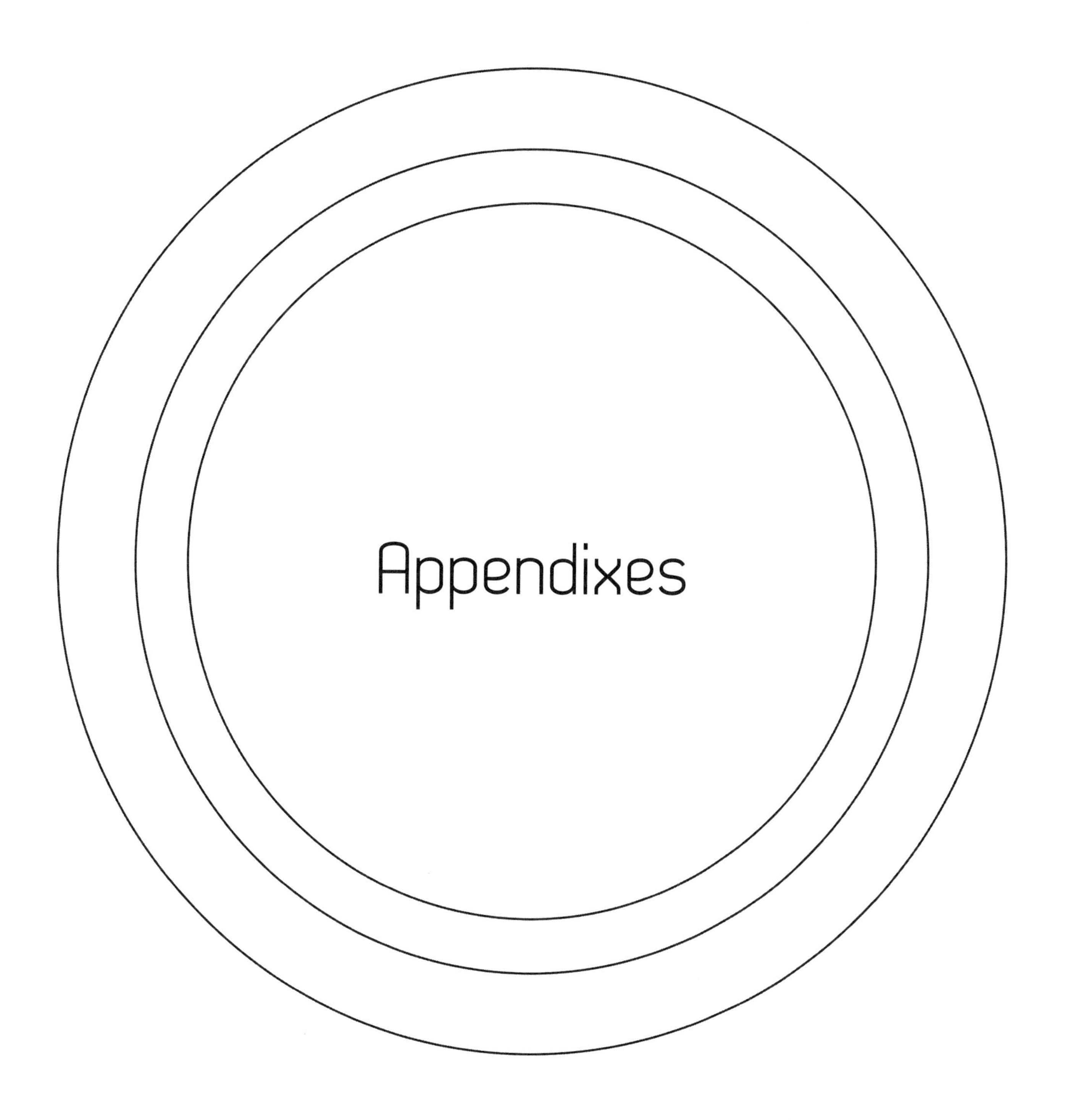

Appendixes

Appendix A. *Next Generation Science Standards*

There is no doubt in the science teaching profession that students need to attain the level of understanding needed to meet proficiency in our agreed-upon set of performance expectations. The *Next Generation Science Standards* (*NGSS*) detail those expectations. Additionally, they enable successful performance by students thanks to three main requirements. Learning materials must have disciplinary core ideas, scientific and engineering practices, and crosscutting concepts deeply embedded. *Earth Science Success* follows the criteria used to judge materials that align well with *NGSS*. Elements of the science and engineering practices, disciplinary core ideas, and crosscutting concepts covered in this book work together to support students in this three-dimensional learning.

Regarding *NGSS* for middle level learners, *Earth Science Success* is focused on the three Earth and space science (ESS) categories: ESS1, ESS2, and ESS3. First, the performance expectations in ESS1: Earth's Place in the Universe helps students develop answers to questions, such as, What is Earth's place in the universe? What makes up our solar system? How can the motion of Earth explain seasons? How do people figure out that the Earth and life on Earth have changed through time?

Second, the performance expectations in ESS2: Earth's Systems helps students formulate an answer to questions such as, How do the materials in and on Earth's crust change over time? How does the movement of tectonic plates impact the surface of Earth? How does water influence weather, circulate in the oceans, and shape Earth's surface? What factors interact and influence weather? How have living things changed the Earth and how have Earth's changing conditions impacted living things?

Finally, the performance expectations in ESS3: Earth and Human Activity helps students generate answers to questions such as, How is the availability of resources related to naturally occurring processes? How can natural hazards be predicted? How do human activities affect Earth systems? Many engineering design lessons and enrichments are offered as well to complement *NGSS* goals.

The lessons in *Earth Science Success* fit together coherently, build on each other, weave different science disciplines, and help students develop proficiency. They also provide grade-appropriate connections to the *Common Core State Standards* in English Language Arts and Literacy in History/Social Studies, Science, and Technical Subjects. One of many examples of this connection is, "following precisely a multistep procedure when carrying out experiments, taking measurements, or performing technical tasks" (CCSS.ELA-Literacy.RST.6-8.3).

This book provides students with firsthand experiences that support authentic engagement in the practice of science. Students are engaged in multiple practices that blend and work together with crosscutting concepts to support students in making sense of both phenomena and solutions. The book uses scientifically accurate and grade-appropriate scientific information, phenomena, and representations to support students' three-dimensional learning. Regarding the *Common Core State Standards*, it provides opportunities for students to express, clarify, justify, interpret, and represent their ideas and respond to peer and teacher feedback orally and in written form.

Additionally, an effort is made to connect instruction to the students' home, neighborhood, community, and culture as often as possible. The book also provides many enrichment opportunities, which are consistent with the learning progression for students with high interest or who have already met the performance expectations.

Appendix B. Electronic Tablet Information

What kind of electronic device(s) should school districts invest in for their students? If you asked early adolescents, they would logically want them all. The realistic answer, however, depends on how students will use the technology for learning.

Different devices provide efficiency for different classroom tasks. Whether it is for creating presentations, writing essays, collaborating with peers, conducting science experiments, taking notes, recording videos, taking photos, reading articles, measuring with electronic probes, or using the internet, the choice of the device depends on the task. Cost, maintenance, delivery, and accessibility all need to be reviewed with a mind on best practice for all disciplines. Educators and IT departments must collaborate with and among each other to ensure effective implementation for all involved. Classroom management becomes of key importance when devices add distraction capabilities, such as games and social media. In that regard, the school devices should be kept as exclusively learning-related tools.

By including current options, the risk of making the book appear dated is real. Regardless, at printing there are currently the following four options:

1. Apple's iPad: The most widespread in school districts across the country. Benefits include easy portability, interactive touch screens, quick boot-up times, and flexibility.

2. Google Chromebooks: Lightweight laptops that run software from the internet instead of a local hard drive. They include built-in access to the Google Apps suite of web-based software.

3. Android's Galaxy, Unobook, and Transformer Pad: These connect tablets to TVs and digital whiteboards in the classroom. Some feature docking stations that bridge devices, as well.

4. KUNO from Curriculum Loft: These feature a strong web filter for the school district to take advantage of, and easy learning management systems.

Important, yet not teaching-related, considerations include rugged screens and cases and lengthy battery life. Student input is vital, along with input from parents and families. As technologies advance and improve, so will learning.

Appendix C. Favorite iPad Apps

- *Notability, Evernote**, and *Paperport Notes:* Fully integrated note-taking apps
- *WeatherBug**: Provides information from classroom weather stations worldwide
- *Explain Everything**: Virtual canvas for screen casting and educational video creation
- *Schoology** and *Showbie:* Fully integrated learning management systems
- *Kahoot!**: Create and play quizzes and surveys
- *Star Walk**: Astronomy and stargazing guide
- *Scan**: Transforms the mobile device into a portable scanner, especially helpful with QR Codes for direct links to sites
- *Spacecraft 3D**: Brings NASA's robotic spacecraft to life in 3D
- *Keynote:* Fully integrated presentation app
- *Book Creator**: Makes e-book publishing easy
- *ComicBook!:* Comic book creator with many features
- *LabTimer:* Stopwatch, counter, and alarm clock
- *NASA**: Huge collection of the latest NASA content
- *The Elements: A Visual Exploration:* Truly the best periodic table app
- *Flashcardlet* and *Quizlet**: Allows students to play and create their own flashcards
- *Google Drive**: Fully integrated learning management system, with opportunities for peer collaboration
- *BioInteractive EarthViewer** and *Earth-Now**: Interactive tools for exploring Earth
- *myHomework Student Planner** and *InClass:* Allows students to plan and manage their due dates

- *Air Server:* Not an app, but allows the teacher to mirror from Mac computers, so iPads can be Air Played on the classroom projector screen
- *SciShow, V Sauce,* and *Kahn Academy:* Not apps, but provide a wealth of video-based science information

* = Android compatible versions of this app are also available.

Appendix D. Six Additional Earth Science Lessons

1. Tic-Tac-Know

This choice-oriented listing of assignments is related to the scientific inquiry process. It is designed to encourage collecting and categorizing of evidence, as well as application, analysis, and extensions of your learning. It uses the educational strategies Bloom's Taxonomy to challenge you as an individual while you learn.

You must select and skillfully complete one activity from each horizontal row on the Tic-Tac-Know Options Board. This will help you, and potentially others, learn more about the scientific inquiry process. Remember to make your work rich in detail, thoughtful, original, and accurate.

FIGURE A.1.

TIC-TAC-KNOW OPTIONS BOARD

TYPE OF INTELLIGENCE	KNOWLEDGE AND COMPREHENSION	APPLICATION	ANALYSIS AND SYNTHESIS
Logical-Mathematical	Make a timeline that shows 25 famous scientists throughout history. Include the date of his or her birth and the dates associated with any important findings.	Design a flow chart that shows the advancements in a particular technological device used for data collection, as it was developed and refined through the years.	Develop and present a *Keynote* that shows a Top-10 list of common mistakes that students could make when conducting an experiment.
Spatial-Artistic	Draw a colored and labeled diagram on a poster, showing at least 10 things that you could do to control all of the variables involved in a typical plant experiment.	Use your iPad camera to complete a collage of photographs of scientific principles found around your home.	Create an information-filled comic book that lists at least 10 things that you could do to test reaction times among teenagers.
Bodily-Kinesthetic	Plan for and involve the entire class in an exercise-related activity, which helps the class remember the steps involved in the scientific inquiry process.	Build a model of an invention you have designed that would help a student count or measure data in a particular experiment.	Prepare and present a skit to the class that shows how to select reliable and credible resources for background research on any topic.

2. Celebrating Mother's Day With a Science Assignment

Can you imagine anything cooler or more fun than celebrating an event by doing a science assignment? I thought not. Here is your opportunity to celebrate someone who has helped to "mother" you.

Many people have played a "mother" role in your life. It could be your own mom, a special Aunt, a family friend, an older sister, a grandmother, a former teacher, a counselor, a neighbor, a friend's mom, a member of your faith community, or a coach.

Mother's Day is celebrated every second Sunday in May. Your task for this assignment is to choose someone special in your life to celebrate in a unique science-related way. Design and create a greeting card for that important mother figure in your life.

Guidelines for the Greeting Card

1. The card must be appropriate, respectful, and reflect the gratitude you feel toward this person.
2. You must use at least 15 science terms from class in the text of your message.
3. The terms must be underlined and used correctly, with enough detail indicating that you know the correct meaning of each.
4. The rough draft will be completed in class, and must be your own original work.
5. The final draft can be electronic or handwritten. Either way, it must be neat and readable.
6. Two colorful images need to be on the card: one on the outside of the card, and one on the inside.
7. The art should match the theme of the card.
8. If your use of the words makes their meaning unclear, you must include a definition key on the back of the card.
9. You must sign the card.
10. The final draft will be shared within your lab group, prior to a check by the teacher.

Sample Ideas

- I look up to you like I'm looking at Polaris.
- You don't pressure me a lot, and compliments come in jet streams.
- Your love works like the greenhouse effect in the way it warms our home.
- My love for you will never erode.
- You act like rising atmospheric pressure because you always bring nice days.
- Your percent human errors are very small.
- As I go through my own rotations and revolutions, spinning through life; you are constantly orbiting me and making sure that I am okay.

3. Time Capsule Activity

Directions

Scientists make predictions based on what they observe about the world around them. A prediction that a scientist makes, explaining a natural event, is sometimes called an "educated guess," for example, "I predict it will be red." (This is different from a *hypothesis*, which is an explanation that can be tested; for example, "It will be blue because red is lower in energy.")

Use your surroundings and your knowledge about the world to make at least three predictions in each of the following categories. These predictions should be targeted to come true between now and the end of this school year.

Your Family and Friends

1.
2.
3. Who will be your best friend at the end of the year?

The United States

1.
2.
3. What leading topic will be in the news at the end of this school year?

The World

1.
2.
3. What particular city might get hit by an earthquake, and what day?

Television and Movies

1.
2.
3. Which actor/actress will win an Oscar for being best in their movie?

Music and Theater

1.
2.
3. Which singer will win a Grammy for being best in music?

Sports

1.
2.
3. Who will win the World Series in October?
4. Who will win the Super Bowl in January?
5. Who will win the NCAA Basketball Tournament in March?

Our School

1.
2.
3. Favorite all-time class at our school?

On the back of this page, trace your hand. Then, write the following sentence to complete your prediction. "I predict that my hand will (grow larger OR remain the same size) by the end of the school year."

4. Earth Science Bingo

Walk around the room and introduce yourself to your classmates. If your classmate is able to answer these Earth science–related questions in a bingo square, write his or her name in the square. Fill in the whole bingo board (not just a row or column) then sit back down in your seat. Each student's name can only appear once on your bingo board. You can use your own name in only one square.

FIGURE A.2.

EARTH SCIENCE BINGO BOARD

I went for a full-day hike in nature this summer.	This summer, I visited a state that has had an earthquake.	I like to look at the stars and planets in the night sky.	I know a scientific term in Latin or Greek.	Autumn is my favorite season.
I like to do experiments and write up detailed lab reports!	I like to sketch and draw what I see.	I read a science-related book this summer.	I have seen a volcano.	I have liked dinosaurs since I was a little kid.
I would like to become an astronaut and go visit Mars.	I like to wonder about things that go on in the world around me.	**Free Space**	I play a musical instrument that uses wind to make sound.	My favorite TV show is the weather portion of the evening news.
I have seen a double rainbow.	I like to take pictures of cool things that I see outside.	I have seen a geyser or visited a hot spring.	I can name Earth's closest star.	I visited a state or national park this summer.
I can name my favorite insect.	I think science is really great!	I can explain why we have seasons on Earth.	I can explain why we see different phases of the Moon.	I can name a famous scientist.

5. Oh, the Science-Related Places You Could Go ... (A Once-Per-Quarter Extra Credit Family Homework Opportunity)

Directions

Go to a science-related place with a parent or guardian to learn as much as you possibly can. Complete the Required Information form below and turn it in by the final week of any quarter. This will earn you extra credit points, if you have no missing assignments during the particular quarter.

Required Information

1. Specifically where did you go (full name and address)?
2. When did you go (must be during the current quarter)?
3. Who went with you?
4. Brief description of the science-related aspects of the event/location:
5. Attach a pamphlet, brochure, or photo from your trip.
6. Student responses to the following:
   a. I learned ...
   b. I most enjoyed ...
7. Parent(s) responses to the following:
   c. I learned ...
   d. I most enjoyed ...
8. Parent signature:

6. Comparing and Contrasting Thinking Maps

Directions

Get to know the people in your class. Complete the Comparing and Contrasting Diagram below, while answering the interview questions with your partner. During the interview, you will discover ways that you and your partner are similar, as well as ways that you are different from each other. Write the findings into the diagram. Be prepared to share several similarities and several differences, as you introduce your partner to the class.

FIGURE A.3.

COMPARING AND CONTRASTING DIAGRAM

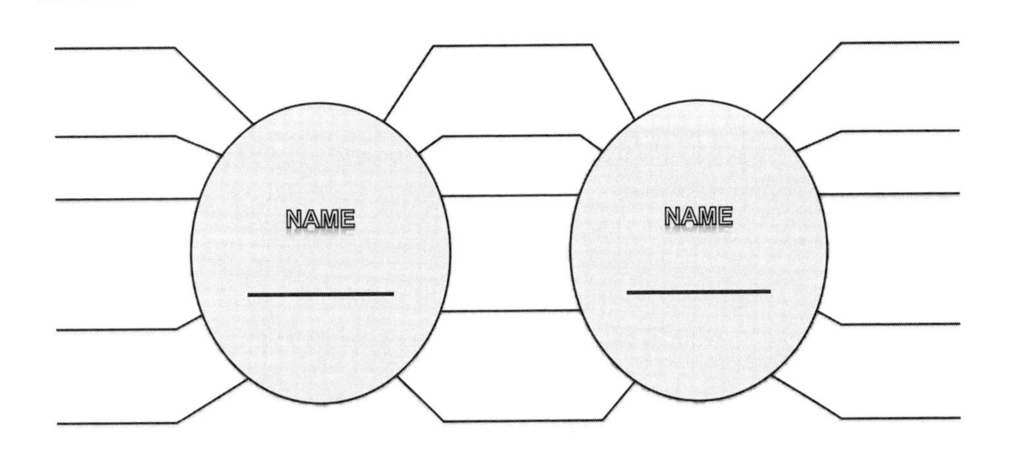

Interview Questions

1. What is your name?
2. What is your favorite month of the year?
3. What is your favorite cartoon character?
4. What is your favorite type of music?
5. What is your favorite plant?
6. What is your favorite form of exercise?
7. Do you prefer winter, spring, summer, or fall?

8. Are you a morning person or a night person?
9. Would you rather text or talk?
10. How many people are in your family?
11. What kinds of pets do you have (names)?
12. What is your favorite way to relax?
13. What is your favorite television commercial?
14. What is your favorite YouTube clip?
15. If you could buy any book or magazine, which would it be?
16. What do you like to spend your money on?
17. What is your favorite vacation spot?
18. What do you do, in order to cope in a healthy way, with frustration or anger?
19. What do you do, in order to cope in a healthy way, with sadness?
20. What career would you like to have when you finish school?

Index

Page numbers printed in **boldface** *type refer to figures or tables.*

Q